ROUTLEDGE LIBRARY EDITIONS:
TRANSPORT ECONOMICS

Volume 13

LABOUR LAW AND OFF-SHORE OIL

LABOUR LAW AND OFF-SHORE OIL

JONATHAN KITCHEN

Routledge
Taylor & Francis Group
LONDON AND NEW YORK

First published in 1977 by Croom Helm Ltd.

This edition first published in 2017
by Routledge
2 Park Square, Milton Park, Abingdon, Oxon OX14 4RN

and by Routledge
605 Third Avenue, New York, NY 10017

Routledge is an imprint of the Taylor & Francis Group, an informa business

British Library Cataloguing in Publication Data
A catalogue record for this book is available from the British Library

ISBN: 978-0-415-78484-9 (Set)
ISBN: 978-1-315-20175-7 (Set) (ebk)
ISBN: 978-1-138-63046-8 (Volume 13) (hbk)
ISBN: 978-1-315-20939-5 (Volume 13) (ebk)

Publisher's Note
The publisher has gone to great lengths to ensure the quality of this reprint but points out that some imperfections in the original copies may be apparent.

Disclaimer
The publisher has made every effort to trace copyright holders and would welcome correspondence from those they have been unable to trace.

LABOUR LAW AND OFF-SHORE OIL

JONATHAN KITCHEN

CROOM HELM LONDON

Croom Helm Ltd.
2-10 St John's Road, London SW11

ISBN 0-85664-395-5

Printed in Great Britain by Biddles Ltd, Guildford, Surrey

CONTENTS

TABLES AND FIGURES

in memory of my father

UK CONTINENTAL SHELF — SPRING 1976

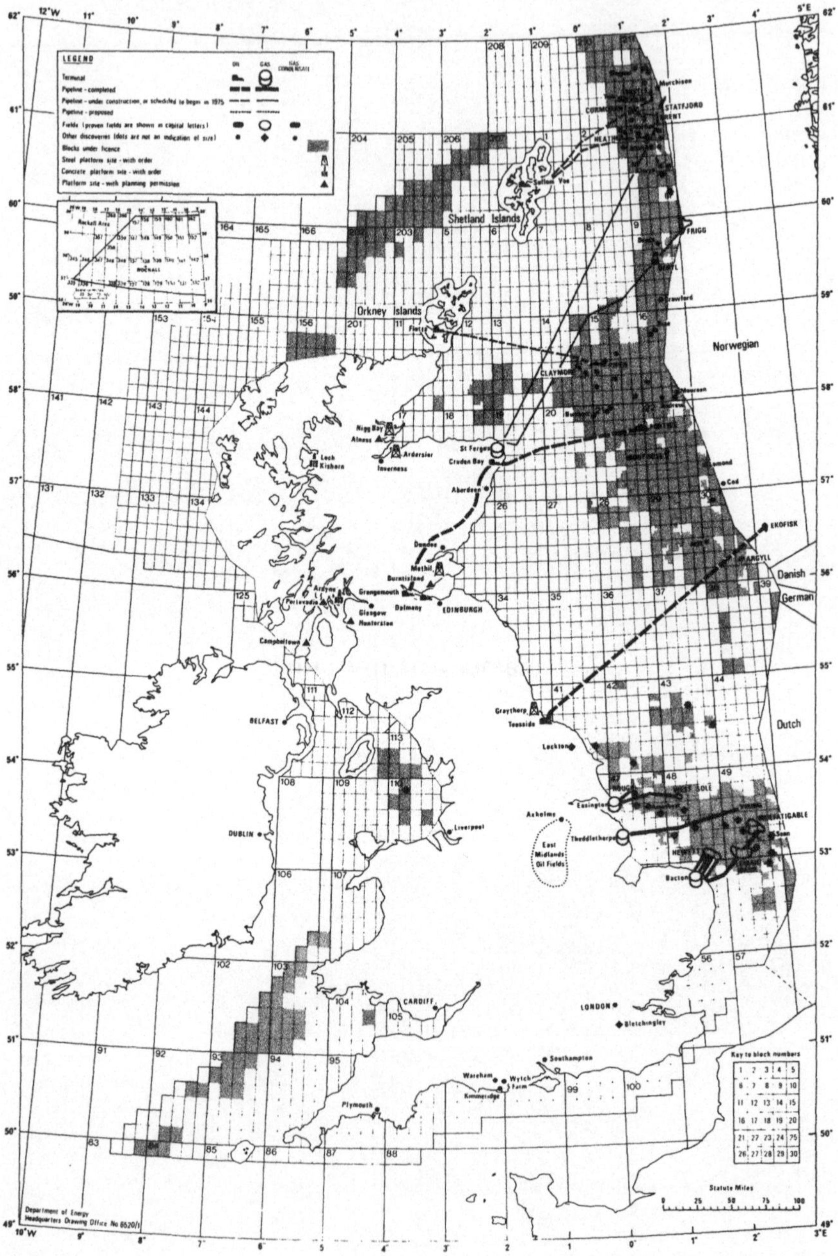

Reproduced by permission of the Controller of Her Majesty's Stationery Office

Based on a map first published in *Development of the oil and gas resources of the United Kingdom 1976*, HMSO, 1976.

PREFACE

> 'I have told the country before that capitalism has its unacceptable face. If you want to see the acceptable face of capitalism, go out to an oil-rig in the North Sea.' (Mr Edward Heath, quoted in *The Observer,* 23 February 1974.)

In the popular mind the principle of equality before the law is sometimes confused with the idea that the law applies to everyone in the same way. But there are some groups of people who are excluded for one reason or another from many of the benefits or burdens of legislation passed by Parliament to regulate or improve behaviour in certain areas of life or to whom such legislation applies in a peculiar way. In the field of employment and labour law those who work offshore on oil-rigs and platforms in the North Sea are such a group. The fact that they work outside Great Britain, but not in another country, has meant that as North Sea oil and gas have become increasingly important to the national economy, domestic politics and international relations, rather special considerations have been applied to the regulation of their employment.

This work therefore attempts to look closely at the way in which labour law applies to such people. More than that, it attempts to look at the way in which the law is applied to working in a dynamic industry new to Britain which operates within severe physical, economic and political constraints. For law does not operate within a vacuum, and labour law cannot be considered in any meaningful way apart from the organisations which are part of the employment process, the policies and attitudes they and their representatives adopt and their purposes and directions. Nor can the responses of the Government to the development of so vital an industry be ignored. Questions of taxation and public participation affect employment prospects, and levels of profit and opportunities change the type of company involved and its behaviour.

But not only is labour law part of a wider genus that reflects society's moods and standards, so its provisions are but one example of the myriad of rules of many different kinds which regulate the employment process. Industrial practice has to be examined wherever possible as an integral part of labour law. For this reason if for no other the study of a small group of workers can be justified: only by getting down to specific examples can the interplay of legal rules, informal practices and social and economic forces be examined effectively.

In fact the study of small groups is equally important for a number of reasons. Although such studies have contributed to the development of the sociology of work, and many that have concentrated on historical aspects or on collective bargaining are vital to the study of industrial relations, few have ventured to cross the narrow line that separates industrial relations from labour law, despite the obvious advantages of so doing. By looking at a defined group of workers in relation to the law the general rules are thrown into relief and their importance is seen in perspective. Furthermore, the pressures which produce legislation and the effect of government intervention can be more readily appreciated.

Of all the groups which might usefully be examined in this way, oil-rig workers merit more attention than most. Firstly, despite the vast outpouring of literature, particularly in international law, on the legal regime for the exploration of the oceans, and the nature, extent and importance of the continental shelf and its limits, virtually no consideration at all has been given to the difficult problems associated with the operation of domestic law on offshore installations, and its application to those who work on board. Secondly, despite the importance of offshore operations and the special legislation which has been passed to regulate these activities and cover the employment of offshore workers, no mention of them is to be found in works on labour law. The reasons are several, but among the most obvious are the fact that those who work on the continental shelf are, like merchant seamen, excluded from virtually all labour legislation, and are only brought under specific Acts by special provisions too complex to be dealt with in general works. Furthermore they are supervised for the most part not by the Department of Employment, but by the Department of Energy. This in turn provides a third reason: without the benefit of protective legislation such as that which applies to employment ashore, those who work on the continental shelf in difficult and dangerous circumstances for large multi-national concerns and who, unlike merchant seamen, do not have strong collective institutions and trade unions to lend them support, are potentially extremely vulnerable in the effort to develop the resources of the North Sea as fast as possible. The very nature of the work as well as the difficulties inherent in the regulatory process means that many legal questions are bound to arise in this connection, and it is hoped that those involved in offshore operations whether as employers or as employees will gain something from this work as well as those interested in the development of labour law and industrial relations in general.

One further advantage of small group studies is that theories normally found in other disciplines can be introduced to highlight certain points. In this respect it should be noted how Erving Goffman's concept of the 'total' institution and his description of the attributes of places in which people are cut off from society for periods of time and lead very separate and ordered existences, has pervaded the approach of government departments and official inquiries alike. Although it is possible to criticise the concept, there can be no doubt as to its effect. Furthermore, in order to attempt to determine what should be classified as 'labour law' for the purposes of a study such as this, drawing upon the notion of a 'career' in the industry and discussing the law that is relevant to each stage in the worker's employment, has to some extent solved both this problem and that of the structure of the study itself.

In consequence, the first chapter discusses the industry, the way it has developed, and the responses of the Government to its increased importance. Chapter 2 outlines the legal regime within which operations on the continental shelf are conducted, and surveys the legislation both of the USA, which was the first western nation to regulate activities on the continental shelf, and of the other countries around the North Sea where offshore operations are also in progress. The possibility of offshore rigs and their crews passing from one part of the North Sea to another means that consideration of the provisions of these other countries might be necessary in some circumstances, although it should be emphasised that it has not been the intention to look at the rules as they apply throughout the North Sea, but to examine in particular those applicable in the British sector. From this point on the book looks at more strictly labour law matters relevant to employment in the British sector of the continental shelf, and passes from consideration of recruitment, training and discrimination to questions of collective bargaining and the form and terms of the contract of employment. Chapter 5 looks at the all-important provisions designed to minimise accidents and promote safety, and Chapter 6 pinpoints some of the principal consequences and associated problems if an employee is involved in an accident out on a rig. Finally Chapter 7 examines the disciplinary rules and the application of the legal provisions relating to dismissal.

This work grew out of part of a much wider inquiry undertaken for the PhD degree at the University of Cambridge. It is therefore my pleasant duty to have to record my thanks firstly to the Master and Fellows of Sidney Sussex College for electing me to an Evan Lewis-Thomas Research Studentship without which it would never have been started, and to the Master and Fellows of Churchill College for electing me to the Bodossakis Fellowship without which it would never have been completed.

In relation to the substance of the work my thanks are due to a great number of people, and it would be impossible to single out by name the numerous officials and members of various companies, trade unions and government departments who have been helpful in countless ways. Of the companies directly concerned, however, I had particularly detailed any helpful discussions with representatives of the Shell organisation, British Petroleum, the Offshore Exploration and Drilling Co., Conoco, and Reading and Bates Ltd. I hope I will be forgiven for singling out J.A.G. Cotton of British Petroleum, G.B. Claydon Esq. of the Department of Energy, Bob Middleton of the North Sea Oil Action Committee, and John Prescott MP for special thanks. Apart from the many friends who helped and sustained me throughout the work I must also thank Dr Patrick Elias of Pembroke College Cambridge, for commenting on an earlier draft, Kit Carson of the Department of Criminology, Edinburgh University, for bringing the work to the attention of the publishers, Professor Howard Williams of Stanford Law School for encouraging and reviewing my interest in the American material, Professor Bill Gould and Professor David Rosenhan, both also of Stanford Law School, for encouragement and support, and above all Dr Paul O'Higgins of Christ's College, Cambridge, whose help and advice throughout all stages of the work has been invaluable and whose kindness will be difficult to repay. Finally, I am deeply indebted to Penny Thomson for typing the manuscript, and for doing much more for me than one could normally ask of any friend, and, of course, to my publisher, David Croom, for his enthusiasm and efficiency which have made the whole process so pleasant.

I have endeavoured to state the legal position as at 1 March 1976. Some amendments were made to the proofs to take account of regulations and legislation published after the text was completed.

California,
July 1976.

1 INDUSTRIAL DEVELOPMENT AND GOVERNMENT RESPONSE

'The rules governing employment, and the way in which they are made and interpreted, cannot be understood apart from the organisations which take part in the process.' (H.A. Clegg, *The System of Industrial Relations in Great Britain,* Blackwell, Oxford, 1970, p.2.)

Offshore oil is now rarely out of the news. Each new development in the process of extracting more and more of this vital energy source from the hazardous waters of the North Sea is meticulously charted in the British press. For Britain, of course, the potential savings to the balance of payments of domestically produced oil, and the prospect that the country might even be in a position some time in the near future to export it, has given the whole subject an importance which makes it relevant to the lives of the whole population. Oil has thus become a part of the domestic political scene as well as a part of international politics. It affects discussions relating to devolution and taxation. It is part of the debate on nationalisation, and its effect on employment, housing, and the order books of British companies is closely scrutinised.

Similar discussions are, of course, taking place in all those countries affected by the recent discovery of offshore oil. The same arguments crop up in Norway, for example, and in Holland and Denmark. In other parts of the world, too, improved technology has made offshore oil important. In Venezuela, Nigeria, Australia, Japan and North America, as well as the traditional Middle-Eastern oil producing nations the same kinds of arguments about oil and its effects take place at one level of government or another.

The person who works on an offshore installation out in the North Sea thus works in an industry which is closely tied to, and affected by, current issues of public policy. Naturally they have an effect on job opportunities, job security, and industrial relations policy. They bear particularly on the way those who work for the industry are treated. How the industry develops, how it is treated by governments, and how it behaves, are therefore all matters which must be examined by the person seeking to understand how the law relating to working offshore operates, how it is relevant, why it takes the form it does, and how it is likely to be used, interpreted and changed.

I. EARLY DEVELOPMENTS[1]

The geographical location of the world's major oil deposits has shaped the petroleum production industry. The fact that the ever-increasing demand for petrol, oil and gas products is located in the major advanced capitalist nations who are insufficiently endowed with the necessary deposits forced the industry to become international almost from its conception.[2] It is now dominated by a small number of huge multi-national corporations operative throughout the world, all of which are concerned, not only to market the available products, but also to discover and exploit further deposits. This search has been given enormous impetus by the obvious political and financial advantage that would result from the deposits being located nearer the markets. Even before the oil crisis was brought about by the Middle-Eastern states in October 1973, attention had been increasingly focused on countries like Alaska, and areas like the North Sea. From the marketing point of view, of course, oil deposits in northern Europe could not be strategically better placed.

It was hardly accident that led the major oil companies to the North Sea, to Great Yarmouth and to Aberdeen. Geological and seismological surveys had, in fact, been undertaken in many areas of the Continent and the United Kingdom for some considerable period of time. Indeed, onshore oil exploration in Britain has been going on for over sixty years, although with limited success.[3] But in 1959 a large natural gas field was discovered in the Dutch province of Gröningen, some twelve miles from the coast. The short distance of this major gas field[4] from the sea redoubled suspicions that the rock formation under the North Sea might be gas bearing. Some seismic survey work had already been done in the late 1950s,[5] but 1963 saw three full survey parties at work in the North Sea.[6] Preliminary surveys were made off the British coast by BRP in the autumn of that year.[7] Actual drilling began in the North Sea in the summer of 1963[8] and towards the end of the year British Petroleum drilled the first (test) well in British territorial waters.[9] Two years later, in October 1965, BP discovered the first North Sea natural gas in the 'West Sole' field 45 miles east of the Humber estuary, and within a year it was found by three other groups off the Norfolk coast: 'Leman Bank' by the Shell/Esso and Gas Council/Amoco partnerships, 'Indefatigable' by the same combinations, and the 'Hewett' field by the Phillips and Arpet groups.[10]

After 1965 the estimates of the possible economic outcome increased steadily.[11] One can, however, point to two early lulls in the development. Before the discovery of gas in the 'West Sole' field the absence of a legal regime for exploration and production offshore prevented the immediate rush that might otherwise have followed the Gröningen discovery. With the division of the continental shelf into 'sectors' under the jurisdiction of the coastal states, and the passage of the Continental Shelf Act 1964 in Britain, and similar legislation in the other countries surrounding the North Sea,[12] this situation was rectified by the subsequent allocation of licences to blocks of the continental shelf giving permission for drilling to take place.[13]

In the United Kingdom sector a more serious setback was experienced after discovery of gas, centring around the negotiations between the discovering companies and the monopolist Gas Council[14] over the purchase price of the gas to be piped ashore.[15] Hints from some sections of the Labour Party, then in office, of moves to nationalise part of the industry, or at least to induce a greater measure of state participation, also acted as further dampeners in this period on the enthusiasm previously shown by the oil companies.[16] In this respect, the companies involved, many of American parentage,[17] had few problems compared to those they have faced in the Middle East. In fact, the discovery of a further supply of energy so close to the shores of Britain might have provoked a sharper reaction, especially as the other forms of energy being used by Britain were in relative national decline. The long-term future of the run-down coal industry was already uncertain, and indeed production has gone down since. The powerful lead that Britain had built uo in the field of nuclear energy showed few signs of being sustained through indecision over which type of nuclear reactor should be used in the future.[18]

It is in fact since 1970 that most of the significant developments in the industry have taken place. With the change of government in Britain the industry 'picked up' tremendously towards the end of the year.[19] Many new companies sought financial involvement.[20] By the end of 1971 further gas finds made it possible to estimate that 90 per cent of gas supplied in the United Kingdom was natural gas from the North Sea, and the discovery of the giant 'Frigg' field in the north in 1972 will mean this figure will certainly remain steady in the future, if not improve.[21] But the most important change was produced by the long-awaited discovery of oil, first in the Norwegian sector of the continental shelf in the 'Ekofisk' field, and then in the

British sector in September 1969 with the 'Montrose' field,[22] and in particular with the discovery of the giant 'Forties' field in November 1970.[23]

II. THE DISCOVERY OF OIL

Although it had long been expected, the discovery of oil changed everything. The political situation in Scotland, for example, was transformed almost overnight, and the regional economy of eastern Scotland would very soon receive a major blood transfusion, even if it has had its price.[24] But the most important effect, of course, has been on the national political scene, and on government policies, through the projected benefits assumed to be forthcoming to the national economy. Immediately, overseas borrowings seemed to have a domestic guarantee attached, and the balance of payments deficit, so long a major government headache, appeared in a different light. In 1973, for example, the UK had an overall balance of payments deficit on her current account of £1,117 million. Oil imports cost the country more than that: £1,281 million. Rapid development of North Sea oil would potentially answer the problem, although, of course, reliance on the development of domestic production immediately placed the developers in an extremely strong position with the Government.

As in the case of the other countries around the North Sea who stood to benefit in a similar fashion, this situation took on a new dimension when the average f.o.b. price of oil from the Middle East was raised by OPEC[25] from about $2 a barrel to over $10 per barrel between September 1972 and mid-1974.[26] The UK balance of payments deficit jumped to £3,838 million in 1974, oil imports costing the country £4,142 million. But not only does a rise in the world price of oil devastate the economy: when combined with the potential threat of a fall at some future date it plays havoc with exploration and production plans, and in particular with the interrelated problem of taxation, placing the companies, and the Government, in an extremely vulnerable position. If the world price of oil rises it automatically becomes economic to develop a certain number of the smaller fields, all other factors being equal. If taxes are introduced the value of some of these fields again becomes questionable, even if the tax is levied in the expectation of the price being at the new high level. If the world market price of crude were then forced down again, the expectations of both the companies and the Government would be dashed, and the effect would be as devastating as the initial increase. Hence the attempts to reach an international agreement on the *minimum* price for crude oil. The important questions which demand attention, therefore, are related to the amount of oil available, the speed with which it can be recovered, and the costs and profits of the companies involved and the governments concerned.

In early 1972 it was estimated that the industry would be able to bring ashore some one million barrels of oil a day by 1975,[27] a figure which now looks as though it was produced in a rush of hopeless overoptimism. But despite increasingly large estimates of the costs of development, the discovery of the major 'Brent' field by the Shell/Esso partnership in July 1971, and the 'Argyll' field by Hamilton Brothers, seemed to make this figure more realistic.[28] Other discoveries followed rapidly. The 'Beryl', 'Thistle', 'Dunlin', 'Auk', and 'Cormorant' fields were also found in 1972, and the 'Piper' and 'Maureen' field in the first half of 1973.[29]

Estimates of the future production consequently rose to between 70 and 100 million metric tons a year by 1980,[30] or 3-4 million barrels a day,[31] more than enough for self-sufficiency. Further developments took place in late 1973 and early 1974 with the discovery of the 'Hutton' field by the Conoco/Gulf/National Coal Board Group in September 1973, the 'Heather' field in November, the discovery of the giant 'Ninian' field by BP/Ranger in January 1974,[32] and by Burmah Oil in February, as well as 'Claymore' by Occidental in June, and smaller fields such as the 'Alwyn' (Total). By April 1974 ten fields had been declared commercial[33] and since then there have been more major finds. BP discovered the 'Magnus' field in July 1974,[34] and announced the discovery of the 'Andrew' field in October.[35] Texaco struck another big field in February 1975,[36] and the NCB/Gulf/Conoco combination announced that the giant 'Statfjord' field extended over from the Norwegian sector.[37] In fact 1975 was an extremely successful year in terms of exploration. Some twenty-two oil discoveries were made in the first nine months of the year, out of a total up to that point of forty-seven. These included the 'Brae' field,[38] and the 'Murchison' field discovered by the NCB/Gulf/Conoco partnership.[39] Other significant finds were made by Amoco[40] and by Hamilton Brothers,[41] as well as a crop of smaller discoveries, for example by Total, and the Transworld, Transocean and Panocean combinations,[42] most of which would need further appraisal work before their commercial value could be estimated.

It should not be thought, however, that the finds are confined to the UK sector of the North Sea, although the most exciting developments are probably taking place there. Important finds have been made in the Norwegian sector, especially the giant 'Ekofisk' oil fields complex in the southern area, the 'Frigg' and 'Heimdal' gas fields further north, and the giant 'Statfjord' oil field, most of which is in the Norwegian sector. Oil has also been discovered in large quantities in the 'Dan' field in the Danish sector, where there have been other smaller finds. Huge quantities of gas have also been found in the Dutch sector, as well as limited showings of oil. There is also a commercial gas field off the south of Eire, the 'Kinsdale Head' field, discovered by Marathon. On the whole, however, the Celtic sea has proved disappointing so far, as has the Atlantic side of Scotland, where oil has been found, but so far not in large quantities.

III. ESTIMATES OF RESERVES AND SPEED OF RECOVERY

As far as the UK sector is concerned, however, the increasing number of discoveries has made it possible to upgrade the estimates of the reserves available for production. In 1974 the Government estimated that proven reserves of some 875 million tons could become a possible 2,950 million tons.[43] By April 1975 this figure has risen by approximately 20 per cent to a proven 1,060 million tons (equivalent to some 7.7 thousand million barrels) and a possible future total of some 3,100 million tons in the existing licensed areas. In fact the Government predicted that reserves could reach 4,100 million tons.[44] In comparative terms, this means that the reserves already discovered in the North Sea exceed those found so far on federal lands off the United States in the Gulf of Mexico, and that the UK oil reserves alone could at some point in the future exceed the current US total reserves.[45] Of course, many of the new discoveries also add to the total reserves of gas, which is now put in the range of twenty-seven to forty-five trillion (10^{12}) cubic feet.[46]

But although estimates of the amount of oil found have been rising, estimates of the rate at which it will be brought ashore have been going down. The five major gas fields in the southern North Sea supplied about 3,500 million cubic feet a day throughout 1974, and existing discoveries should support production of some 5-6,000 million cubic feet a day well into the 1980s and possibly beyond. The 'Brent' oil and gas field alone should yield up to 1,000 million. By contrast, it was originally expected that some 25 million tons of oil would be brought ashore in 1975. By 1974 this estimate had slumped to 5 million, and by

April 1975 was down to somewhere between one and two.[47]

Oil from the British sector was eventually landed. The first oil was carried from the Hamilton Brothers' 'Argyll' field (one of the smallest commercial fields) by tanker and landed on 17 June 1975.[48] On 3 November 1975 the Queen inaugurated production from the 'Forties' field[49] with oil flowing through a pipe-line to Grangemouth at the rate of about 30,000 barrels a day, but expected to rise steadily to about 140,000 in March 1976 and 400,000 at the beginning of 1977, equivalent to nearly 20 per cent of the country's current demand. Meanwhile oil from Norway's 'Ekofisk' field was flowing into Teeside at the rate of nearly 300,000 barrels a day. But in December two anchors from the 'Auk' field's single point mooring buoy broke, delaying production from this field until February 1976,[50] and on 5 December Mobil's 480 ft high steel loading tower for the 'Beryl' field broke free from its foundations on the sea bed and began to drift towards Norway. Although it was recovered, production was expected to be delayed at least six months.[51] There have been other delays too. A tower crane collapsed, putting back the date of production from the 'Brent' field,[52] and the 'Piper' field has been delayed. Production from the 'Frigg' gas field will now be up to two years behind schedule because the production platform was sunk about three kilometers out of place when the flotation equipment failed, causing expensive damage.[53] Nevertheless, several fields are now expected to be producing by the end of 1976, and still more by the end of 1977.

However, there does appear to be a general slowing down in the pace of exploration and production. Exploration activity reached a high peak in 1974 when the total number of rigs used in the British sector rose from twenty-five in 1973 to thirty-nine, and the maximum in use at any one time rose from eighteen to twenty-nine, resulting in an increase in the total time spent in the area by all rigs from 160 rig months to 294 rig months. Between them they drilled or began a total of 100 exploration and appraisal wells. The Government expected this increase to be continued with more rigs being brought into use, and forecast a maximum number of forty-five and a maximum at any one time of between thirty-five and forty.[54] Although their success rate was good, the average number of rigs in operation was only twenty-eight, and the number of rig months of operations was only expected to reach 300.[55] The projections for 1976 have had to be cut back by about one-third, as have the projections for additional orders for production platforms. Work is thus proceeding just as fast, but the rate at which it has been increasing, and was expected to increase, has fallen off with a number of important consequences, for example, a number of rigs have been out of work.

But it is the causes of the slow-down which are really important, and these, of course, are mainly to do with costs and profits. Even when oil was first discovered, there was some apprehension about this high-cost, high-risk venture, and indeed there was some initial criticism that not enough notice was being taken by the Government and by British industry of the enormous potential involved in the shift of emphasis from Great Yarmouth to Aberdeen.[56]

IV. COSTS AND PROFITS

The initial cause of these attitudes was probably the sheer size of the sums of money needed to bring oil ashore from an offshore find, as well as the general sluggishness of British industry to re-tool and adapt to the highly complex technical demands being made by what then appeared to be a high-risk business. Even at the beginning it was about five times as expensive to carry out all the development drilling and other work needed to exploit a commercial find offshore as it was to do similar work on shore.[57] British Petroleum had to raise an initial loan of no less than £360 million to pay for bringing its 'Forties' field into production,[58] and in 1972 it was estimated that the industry might have to spend a total of £2,000 million in order to develop a production of only 2 million barrels a day, equivalent to about 100 million tons per annum.[59] The total *world* demand of the petroleum industry for capital in the 1970s was then put at the 'median figure' of $4,000 billion.[60]

To give some idea of the way in which costs have escalated, in 1972 Sir David Barran of Shell Transport and Trading forecast that the necessary capital investment for exploration and production in the British sector of the North Sea alone would be £1,500 million spread over the next ten years, and that a further £1,000 million would be needed to cover the operating expenses throughout the decade.[61] He also broke down the likely development cost of a northern oil field into: £100 million for platform structures and production facilities; £70 million for drilling costs; £65 million for offshore trunk and service lines; and £15 million for shore-based facilities. Deep water brings a steep rise in investment costs. The investment cost per barrel of oil per day of production is over £400 off the Nigerian coast, over £1,000 in the Gulf of Mexico and in deep water off Scotland anywhere between £2,000 and £3,500.

But these figures, which must be seen as underestimates, have all had to be revised in the light of current inflation rates. Some operators have apparently put the cost escalation of some items and services as high as 400 per cent since 1968.[62] The daily hiring rate for a derrick barge has gone from $17,000 to $70,000; for semi-submersible drilling rigs the increase has been from $14,000 to a peak of over $60,000, although this price has fallen back as a surplus of rigs has developed with the cut-back in the rate of expansion and an increase in the numbers being delivered from building yards. Capital costs have gone the same way. In mid-1973 the cost of developing a medium-size offshore field was put at £130 million. By mid-1974 this had risen to £240 million.[63] In the middle of 1974 Shell admitted that the cost of four production platforms alone for the 'Brent' field would now be about £200 million. These figures will by now have risen again further, and since 1972 the cost of drilling one of the 700 or so exploration, appraisal, and production wells drilled since 1964, at a cost of between £500,000 in the southern part of the North Sea, and £1 million in the northern sector, has almost doubled. The cost of developing the 'Forties' field has now risen to some £700 million,[64] and that of the 'Thistle' field has reportedly tripled in a year.[65] Developing the 'Ninian' field will probably now cost some $3,000 million, making for some of the world's most expensive oil.[66]

On the other hand, the profits of the oil companies have generally reflected this high level of investment. Five major US oil companies had total net earnings for the first half of 1972, that is before the oil crisis, of $1,870 million.[67] The 1972/3 edition of *The Times 1000*[68] placed British Petroleum as Europe's top company by size of turnover, for the second year running, with annual sales of £3,153,300,000 and profit, before interest and tax, of £738,500,000. Royal Dutch Shell ranked second with annual sales of £3,018,600,000, and Shell Transport and Trading third with sales of £2,923,096,000. (It was also the second largest profit-maker with £472,562,000.) British Petroleum, Shell Transport, and Esso Petroleum were respectively first, second and fourth in the table of biggest profit rises during the year. British Petroleum had shown an increase in profit of £272,000,000 (58.3 per cent), Shell Transport of £87,633,000 (22.8 per cent), and Esso of £20,041,000 (169 per cent). Smaller companies did not do so well, in many cases because of rising prices.[69]

Following the rise in the price of oil most of the European oil companies made even bigger profits. BP and Shell Transport and Trading were respectively first and second in *The Times 1000* for the 1975[70] list of top UK companies by turnover with sales at the end of the 1974 accounting year of £4,511,900,000 and £4,232,292,000 respectively. BP made £411,900,000 profit (up 53.1 per cent[71] from the year before), and Shell managed £341,999,000 (up 71.6 per cent).

Nearly all the oil companies rose in the league table,[72] and there were seven oil companies in the top twenty American companies by sales. The equivalent figures at the end of the 1974 accounting period were even bigger (although costs appeared to be rising faster than income[73]), but they will not be as high for 1975.[74]

V. GOVERNMENT RESPONSES

A. Aids to Industry

The response of the UK Government has taken two major directions. Firstly it has been concerned to make sure that British industry gains some of the benefits from spin-off investment and economic development. Secondly, since the Labour Party resumed office in 1974 it has made certain demands for an increased slice in the profits of the oil companies from North Sea oil.

Right at the start of the offshore boom, the Government made repeated attempts to ensure that the investment was not all American, and that British companies were involved. Initially this policy was pursued hesitantly because the economic climate was what might be described as 'cautious', and there was some scepticism as to the possible benefits. But there can be little doubt that the Government was also conscious of the effects of the developments on regions with high unemployment,[75] and more recently the concern has shifted to ensure that the whole of British industry should profit from the opportunities. This gradual change in government thinking has been reflected in legislation.

Under the 1964-70 Labour administration Parliament passed a number of Acts relevant to the offshore industry. Apart from those providing for participation by nationalised industries,[76] Part One of the Industrial Development Act 1966, gave the Board of Trade the power to make a grant to anyone carrying on a business in Great Britain towards capital expenditure incurred in providing new machinery or plant for use in Great Britain.[77] This included machinery intended for use in 'designated areas' of the continental shelf,[78] for 'searching for. . . or extracting or getting. . .oil. . .or natural gas'.[79] The Conservative Government of 1970-4 made many changes in these policies,[80] and the Investment and Building Grants Act 1971 broadly speaking put an end to grants under the Industrial Development Act 1966. But more help was directed to private industry, which is still eligible to receive enormous aids to investment under the Mineral Exploration and Investment Grants Act 1972, and the Industry Act 1972.

Although the Mineral Exploration and Investment Grants Act is designed to operate only in relation to exploration of various non-ferrous metal ores, barium minerals, fluorspar and potash,[81] and any such extension of the offshore industry has yet to make itself manifest,[82] section 1 (1) gives the Secretary of State power to make available sums of money not exceeding £25 million:

> . . .by way of contribution towards any expenditure incurred in searching for, or in discovering and testing, mineral deposits in Great Britain or on the sea bed and subsoil of the territorial waters of the United Kingdom or in that of any area for the time being designated under section 1 (7) of the Continental Shelf Act 1964.

The grant of money can be made 'on such terms as (the Secretary of State) thinks fit',[83] so that it can be either an outright grant with no obligation to repay, or a loan, but the grant cannot exceed 35 per cent of the total cost of the venture.

The Industry Act 1972 provides for a different form of aid, and is designed to help companies to actually obtain equipment necessary to explore the continental shelf for oil. Section 10 of the Act provided that the Secretary of State for Industry may give a guarantee for any company incorporated under the law of any part of the United Kingdom of up to £1 billion for the construction of mobile offshore installations.[84] Section 11 allows the Secretary of State to make a grant 'subject to such conditions as he thinks fit' to persons who have entered into contracts to construct a ship or mobile offshore installation in the United Kingdom and to equip it. The philosophy of the Government was made very clear by the then Secretary of State himself at the Conservative Party Conference at Blackpool in October 1972. Mr John Davies is reported as saying:

> By 1980 we expect that at least half our oil needs will be met from production round our own shores. The Government means in every way to help industry catch this tide of success through the Industry Act, which Mr Chataway administers; by consultancy assistance; and by ensuring that oil companies granted North Sea licences give full and fair opportunities to British industry.[85]

In pursuit of this policy, and following the criticisms made in the IMEG Report, which was commissioned by the Government to examine the possibilities for British Industry in relation to the North Sea developments, an 'Offshore Supplies Office' was set up within the then Department of Trade and Industry to use the powers under the Act.[86] At the same time a 'Scottish Petroleum Office' was formed in Glasgow as part of the Scottish Industrial Development Office,[87] and every effort was made to involve British industry.[88] At the end of 1973 it was reported that there had been an increase in this involvement[89] and during Parliament's Christmas recess a government reshuffle produced a Department of Energy to supervise all activities relating to energy.[90]

Nevertheless the benefits to the oil industry have been slow in forthcoming. Problems in the domestic steel industry have meant that the British Steel Corporation has not been able to produce enough undersea pipeline (although it has produced about 90 per cent of the onshore pipes). British manufacturers have been producing small parts, but few of them have taken the plunge into producing supply vessels, or oil-rigs themselves.[91] Some senior industrialists thought that a delay would help British industry,[92] a viewpoint not shared by the oil industry and government departments alike. Indeed, imports of equipment for the North Sea have been having an adverse effect on Britain's balance of payments. The Government has now stepped up its attempt to direct the benefits of expansion to British companies, securing a guarantee of fair treatment from the organisations representing the major oil companies in an endeavour to raise the level of orders going to British companies from its present level of about 45 per cent.[93] In addition, the passage of the Offshore Petroleum Development (Scotland) Act 1975 should make it easier to construct the giant production platforms, several of which have now been sunk into place, in Britain instead of their being built abroad.[94]

B. Taxation

Perhaps the major problem, however, arises from taxation and government participation. The Labour Government announced in July 1974 its intention to increase public participation in the benefits of offshore oil through a public corporation, the British National Oil Corporation; through alteration of the licences for offshore exploration and production; and through a Petroleum Revenue Tax[95] which was prompted by the 'windfall' profits of the oil companies from the rise in crude oil prices.[96] But the rate at which it was proposed to levy the tax was not immediately revealed, causing a substantial amount of disquiet amongst potential investors and which spread through the industry.[97] The Government was forced to guarantee a number of smaller firms involved in exploration and the Bank of England acquired part of the Burmah Oil stake in the North Sea as a result of the company's financial problems.[98] Several major oil companies threatened to reduce their involvement in the North Sea or delayed investment until the argument with the Government was cleared up.[99] When it was, it was hailed as a victory for the companies,[100] and share prices rose with the news.[101]

The root of this argument has been the inability to assess the size of the reserves being discovered. The true extent of the size of the fields has been hard to determine, and the companies themselves, with an interest in keeping the size secret until after the tax was fixed, have been the only real source of information. They argued that the tax would make the smaller fields, already only 'marginally' profitable, uneconomic.[102] The estimates thus ranged from the pessimistic to talk of self-sufficiency. Constant updating of the amount of proven reserves continues, the rising figures confirming the predictions of Professor Peter Odell[103] that there is much more oil under the North Sea than everyone thinks at present.

The Petroleum Revenue Tax, to be levied under Part One of the Oil Taxation Act 1975, was therefore the result of a number of compromises. Thus the Act, which provides for the tax to be levied at the rate of 45 per cent on the profits from oil won in the UK sector,[104] calculated on a formula basis[105] with special rules for valuation instead of by adjustment of the commercial profits, exempts existing North Sea gas fields,[106] and provides for certain safeguards to protect marginal fields. These include an oil allowance for each field of one million long tons per year which would be free of the tax,[107] and a provision whereby the tax would be cancelled altogether for a field on which the return in any one year does not exceed 30 per cent of the accumulated capital expenditure, with a tapering off provision above 30 per cent.[108] Nevertheless, it is expected that the Government will take on average about 70 per cent of the profits from each field (more for a larger field) when royalty payments and corporation tax are also taken into account.[109] This figure is roughly comparable to that of the Norwegian Government's take in its sector, where the oil tax is seen by the industry as having roughly the same effect as PRT, but being rather harsher on cash-flow in the early and late years of production. In fact, it has been estimated that the Norwegian Government will take about 77 per cent of the profits from the 'Ekofisk' field, measured over the life of the field.[110]

C. Public Participation

It is not therefore very surprising that the North Sea does not appear as rosy an investment for the oil companies as it did once. It seems as though some of the finds in the UK sector may not achieve the 25 per cent discounted cash-flow on investment which is regarded by many companies as the minimum necessary to retain their interest, although much depends on the maintenance of a continued high world price for oil. In addition, the industry has been faced with the Labour Government's plans for increased direct public participation in offshore oil activities.

In fact, direct state participation is not new. In the early years of the offshore industry in the North Sea, Parliament passed the National Coal Board (Additional Powers) Act 1966, which gave the Coal Board powers to 'search and bore for and get petroleum',[111] and the Gas Act 1965 was drafted in sufficiently wide terms to give the Gas Council similar authority.[112] Both these nationalised industries have been involved in the North Sea ever since, and the NCB was particularly successful in its partnership with Gulf and Conoco. The reaction of the then Labour Government was somewhat muted, for although priority was given in the third allocation of licences for blocks of the continental shelf for exploration production to consortia containing the NCB or the Gas Council state participation was described as 'an increasingly haphazard affair'.[113] BP, of course, was one of the foremost companies involved, but the Labour Party group that reported on nationalisation made little immediate headway, and although the Government had decided toward the end of its period of office that all west coast licences[114] should be conditional on 50 per cent state participation, this was naturally shelved when the Conservative Party took power in June 1970.

But the new Government did not end the involvement of the nationalised industries. Indeed when the Gas Act 1965 was repealed and replaced with a similarly entitled Act of 1972 that completely restructured the industry, the new Gas Corporation was specifically given the power to 'search and bore for and get natural gas' and to 'bore for and get petroleum found in the form of crude oil in the course of searching for or getting natural gas'.[115]

Nor should it be thought that state participation in the North Sea is confined to Britain. The Dutch Government have always had an optional 40 per cent interest in production licences, and the Norwegian Government have been involved too, initially through a company called

Norsk Hydro, originally a German company but acquired during the war, and more recently through a state oil company created in 1972 to take over the state's interests in offshore oil and gas. In this respect the creation of the British National Oil Corporation (BNOC) under the Petroleum and Submarine Pipe-lines Act 1975, which gives effect to the rest of the Government's proposals for increased participation, is a close parallel.

The BNOC, which officially began operations at the beginning of 1976,[116] has been the centre of controversy from its conception. Headed by Lord Kearton, a former Chairman of Courtaulds Ltd, the Deputy-Chairman is Lord Balogh, who resigned as Minister of State for Energy and has been the principal target of attack by the opponents of participation in the industry.[117] Under the Act, the Corporation has been given extremely wide powers, for example the power 'to search for and get petroleum existing in its natural state in strata in any part of the world',[118] and to:

> . . .do anything required for the purpose of giving effect to agreements entered into by the Secretary of State with a view to securing participation by the Government of the United Kingdom, or by the Corporation or any other body on behalf of the Government, in activities connected with petroleum beneath controlled waters.[119]

By virtue of section 13 the BNOC immediately took over all the North Sea interests of the NCB by the transfer of the shares of NCB (Exploration) Ltd to the Corporation at their face value.[120] This immediately gave it an interest in several commercial fields, including 'Thistle', 'Dunlin', 'Hutton', 'Brae' and much of the British extension of the 'Statfjord' field, through the NCB's activities in partnership with Gulf Oil and Conoco. In fact the NCB also had interests in a number of gas fields as well, including a 50 per cent share in the 'Viking' field.[121] It should be noted too that the BNOC is exempt from Petroleum Revenue Tax.[122]

Perhaps equally controversial has been BNOC's acquisition of the North Sea interests of Burmah Oil. Burmah found itself in financial difficulties at the beginning of 1975, and had to agree, not only to the principle of 51 per cent government participation in its North Sea finds, but also to pledge its 21.6 per cent shareholding in BP to the Bank of England in return for guarantees from the Bank that would enable it to continue in business. The Bank subsequently bought the shares for

£179 million, giving the Government a potential majority holding in BP.[123] Indeed, the value of the shares then rose, causing some considerable resentment, on top of which the Government offered to buy all of Burmah's stake in the North Sea.[124] The BNOC subsequently paid Burmah some £40 million 'on account' in respect of the company's interest in the 'Ninian' field, with negotiations continuing on Burmah's share of 'Thistle'.[125] In pursuit of the Government's policy the Corporation would like to take a controlling interest in as many fields as possible, but so far only a small number of companies have agreed, even in principle, to a BNOC take-over. In each case even these agreements have been secured only where there has been some prior link with the Government, as in the case of BP and the BNOC's partners Gulf Oil and Conoco,[126] or where approval has been necessary for a change in ownership of the licence, as is the case with Chevron (a subsidiary of Standard Oil of California),[127] or where the company has had to turn to the Government for financial assistance, as was the case, for example, with Ranger Oil.[128]

Offshore oil has thus rapidly become a central issue of government policy. It will undoubtedly provide an enormous amout of wealth for the country, but this is dependent on the ability of the Government to walk the narrow line that separates reaping the rewards through taxation and participation on the one hand, and doing so to the extent that the companies involved lose interest in the activities. Such decisions are of great importance from the point of view of the person working for the industry offshore. There has been a gradual shift of emphasis from exploration to production, and new kinds of jobs, with different prospects, appear as a result. Government participation will mean that attitudes to labour may undergo certain changes, and there is no doubt that increased focus on offshore activities has meant more regulation and a growing concern over matters such as safety. This too has been reflected in legislation.

2 THE LEGAL BACKGROUND TO OFFSHORE OPERATIONS

'What willing allegiance can a man owe to a canon of obligation which is not even conceived in such form as to be understood?' (Lord Radcliffe, 'Some Reflections on Law and Lawyers', 10 C.L.J. 361 at p.368.)

It is obvious that once exploration for oil and gas began in waters off the coast that some legal arrangements should be forthcoming to establish the extent of British 'rights' over the deposits, and to regulate the activities of mining and exploration, and indeed these have had to be made in all countries where the petroleum industry has moved offshore. In addition, the physical presence of persons working on the continental shelf raises questions relating to their subjection to the law of the coastal state. The discussion thus falls naturally into two parts.

I. THE LEGAL REGIME FOR OFFSHORE DRILLING

A. International Law

The legal regime surrounding offshore drilling is of comparatively recent origin. It dates from the 1958 United Nations Conference on the Law of the Sea held in Geneva.[1] Three conventions emerged from the conference: those of the Territorial Sea and the Contiguous Zone; The General Regime of the High Seas; and the Convention on the Continental Shelf. If is not the intention here to provide a detailed account of these conventions, though they remain important as background to the legislation passed as a result. Nor is it intended to review or discuss the volumes of international law literature that have ensued. This section serves only to put the later discussion in perspective.

The UK ratified the Convention on the Continental Shelf on 11 May 1964,[2] being the twentieth nation to do so, and completing the number required to bring it into force. Although the Continental Shelf Act 1964 supposedly enables 'effect to be given to certain provisions of the Convention on the High Seas',[3] it is the Continental Shelf Convention which remains important here. Some parts of it must be examined, especially as the 1964 Act nowhere defines the term 'continental shelf'.

Article 1 of the convention provides that:

For the purposes of these articles the term 'continental shelf' is used as referring

(a) to the sea bed and subsoil of the submarine areas adjacent to the coast but outside the area of the territorial sea, to a depth of 200 metres or, beyond that limit, to where the depth of the superjacent waters admits of the exploitation of the natural resources of the said areas;
(b) to the sea bed and subsoil of submarine areas adjacent to the coasts of islands.

This definition has the major fault of permitting 'creeping jurisdiction' where it is possible to investigate the sea bed and extract minerals from depths greater than 200 metres.[4] In fact extraction is possible in considerably greater depths. In October 1974 the drillship 'Glomar Challenger' drilled a well in 3,737 feet of water 125 miles off Lofoten Island.[5] Any redefinition that would match the technology would thus stretch the limits of the continental shelf well out into the Atlantic and into areas which would offend traditional notions of rights to the sea bed and the high seas.

This problem has led to long arguments at the Third UN Conference on the Law of the Sea[6] and is, of course, tied up with the limits of territorial waters, exclusive mining rights and an international regime for the ocean floor.[7] The conference has now held some four sessions since it began in New York in late 1973, and the various committees have now met in Caracas in 1974, in Geneva in 1975, and, as no agreement has so far been reached and the 1958 convention remains in force, will reconvene again in 1976. Under the 1958 convention, however, the whole of the North Sea, apart from the 'Norwegian Trench' or 'Trough' is within the limits of Article 1, and has therefore been divided up along median lines agreed between the various surrounding countries, in a series of bilateral treaties, into sectors appertaining to each one.[8] The strange shape of the boundaries between the West German and Dutch and Danish sectors reflects the fact that West Germany was not a party to the convention, and that the boundaries were indeed only fixed between the three countries in January 1971 after the matter had been referred to the International Court of Justice.[9] Other boundaries have also given rise to a number of problems. Thus agreement has had to be reached over those fields which straddle the median lines between two sectors, such as the 'Frigg' field,[10] as well as over the laying of pipelines that pass from one sector to another.

Although the United Kingdom has reached agreement in principle over the median line with the Belgian Government, negotiations are still proceeding with the Irish Government over the median line in the Celtic Sea, and with the Norwegian Government for the area north of 61° 44' 12", which is where the present line ends. Discussions will also have to include the Danish authorities because of the Faeroe Islands, and possibly Iceland as well.

Indeed, the complications caused by islands brings into focus two traditional debates,[11] one between the United Kingdom and Ireland over the possession of Rockall,[12] an island (or rock) which might give the possessor rights over an enormous area of the sea bed were the territorial limits extended to 200 miles, and the other between the United Kingdom and France, ostensibly over the Channel Islands but in reality over the median line between the two countries, which manifested itself in the refusal of the French Government to sign the convention at the time of the conference. In fact, although the islands are to be left within the French sector, no median line has yet been agreed between the two countries in respect of rights in the channel. Indeed, although the dispute between the two governments has been submitted to arbitration by an international tribunal,[13] the issue has led to high level diplomatic arguments[14] because the so-called 'Western Approaches' (an area at the west end of the Channel) are supposedly rich in oil deposits. Indeed two large potentially oil-bearing structures have been discovered there by seismic survey,[15] and division along the median line would give Britain the major share. It is nevertheless obvious that the industry is not too concerned about this area at present, although the French Government awarded permits to drill in an area to the south in April 1975. Firstly, development has been concentrated on the proven areas of the North Sea and is likely to do so until the economic climate improves. Secondly, there is no doubt that installations in the Channel itself would create a very dangerous hazard to shipping in what are already the most crowded straits of the world. To permit commercial drilling in this area would cause considerable problems.[16]

The other interesting point is the special position of the Isle of Man. The Isle of Man, though it has territorial waters, has no continental shelf. The area has been subsumed in the continental shelf of the UK in return for a *per capita* payment to the Manx Government in respect of any oil produced in any region of the UK continental shelf. In fact the Manx Government has done very well out of this arrangement as the area that would have been the continental shelf of the Isle of Man has so far proved unproductive!

Article 2 of the convention clarifies the authority of the coastal state over the continental shelf. It provides:

> (1) The coastal state exercises over the continental shelf sovereign rights for the purpose of exploring it and exploiting its natural resources.
> (2) The rights referred to in para. 1 of this article are exclusive in the sense that if the coastal state does not explore the continental shelf or exploit its natural resources, no-one may undertake these activities, or make a claim to the continental shelf, without the express consent of the coastal state.[17]

The other provisions of importance here are all to be found in Article 5, parts of which read as follows:

> (2) . . .the coastal state is entitled to construct and maintain or operate on the continental shelf installations and other devices necessary for its exploration and the exploitation of its natural resources, and to establish safety zones around such installations and devices, and to take in those zones measures necessary for their protection.[18]
> (4) Such installations and devices, though under the jurisdiction of the coastal state do not possess the status of islands. They have no territorial sea of their own, and their presence does not affect the territorial sea of the coastal state.

The somewhat uncertain status of offshore installations is immediately apparent from these provisions. As they do not have the legal status of islands, and in many cases are not registerable as ships,[19] they fall outside the accepted legal categories of territorial extension to which the law of the state claiming ownership or proferring nationality would normally apply. Though the convention acknowledges the sovereign rights of the coastal state over the sea bed, and in a sense confers jurisdiction on the coastal state in respect of installations placed on the sea bottom or maintained on the sea above, such an international accord needs translation into domestic law in order to take legal effect in a substantive manner. Before turning to look in detail at the way this has been done in the United Kingdom, it may be useful to outline the approaches taken by some other western nations.

B. Domestic Legal Frameworks for Offshore Development

1. The United States

Most countries with offshore resources capable of development have legislation of one kind or another in line with the international convention. That of the United States, however, preceded it. Oil was first produced from an offshore structure, a pier off the coast near Santa Barbara, in 1897. Following the Presidential Proclamation issued by President Truman in 1945, laying claim to the resources of the continental shelf,[20] the first well drilled from a platform out of sight of land was completed in 1947 in the Gulf of Mexico off Louisiana. In 1953 President Eisenhower signed into law two pieces of legislation, the Submerged Lands Act,[21] and the Outer Continental Shelf Lands Act,[22] designed to provide a framework for offshore activities, and to put an end to the so-called 'Tidelands' dispute in which the Federal government was embroiled with a number of state governments over ownership of offshore minerals.[23]

Under the Submerged Lands Act

> . . .title to and ownership of the lands beneath navigable waters within the boundaries of the respective States, and the natural resources within such lands and waters, and the right and power to manage, administer, lease, develop, and use the said lands and natural resources all in accordance with applicable State law. . .

was recognised, confirmed and vested in the states.[24] In respect of offshore exploration and production within the territorial limits of the states[25] the legal regime is therefore that laid down by the state in question, and the law applicable to events on offshore rigs or to employment on such a rig is the law applicable in the state. This can, of course, include Federal law.[26] The Federal government also expressly retains its rights and powers of regulation and control of the areas for the constitutional purposes of commerce, navigation, national defence and international affairs, but these do not include rights of ownership, management, administration, use and development of the lands and the natural resources.[27]

Although the Submerged Lands Act specifically reserves to the United States rights in respect of the continental shelf beyond the seaward boundaries of the states,[28] Federal jurisdiction and control are regarded as flowing principally from the Outer Continental Shelf Lands Act (O.C.S.L.A.) which provides for Federal administration of these areas.[29] Under this Act the Secretary of the Interior is authorised to grant oil and gas leases to areas of the outer continental shelf under a system of competitive bidding.[30] The Act lays down provisions relating to royalty payments[31] and gives the Secretary of the Interior power to make regulations governing the bidding system,[32] to include such terms and provisions in the leases as he thinks fit,[33] and in general empowers him to prescribe such rules and regulations as may be necessary to carry out the provisions of the Act relating to leasing,[34] which is done under the supervision of the Bureau of Land Management. The regulations in turn give the Department general control over all offshore operations, which control is in practice exercised through the US Geological Survey,[35] whose Area Oil and Gas Supervisor issues the relevant permits for exploration and drilling activities, although, as will be seen, many other agencies are also involved. These regulations contain many provisions touching on work on offshore rigs. For example, the supervisor has the power to approve the design and features of all 'platforms, fixed structures and artificial islands',[36] and orders issued to licensees under the regulations cover such matters as training of personnel and safety procedures as well as the marking of all installations.[37]

In relation to the law to be applied on the outer continental shelf, section 4 (a) (1) of the Act provides that:[38]

> The Constitution and laws and civil and political jurisdiction of the United States are extended to the subsoil and sea bed of the outer Continental Shelf and to all artificial islands and fixed structures which may be erected thereon for the purpose of exploring for, developing, removing and transporting resources therefrom, to the same extent as if the outer Continental Shelf were an area of exclusive Federal jurisdiction located within a State. . .

Under the Act, as amended by the Deepwater Port Act of 1974,[39] 'the civil and criminal laws of each adjacent State now in effect or hereafter adopted, amended, or repealed. . .' are declared to be the law of the United States for application to the sea bed and subsoil of the outer continental shelf and artificial islands and fixed structures on it, but only to the extent that the state laws 'are applicable and not inconsistent with this Act or with other Federal laws and regulations of the Secretary now in effect or hereafter adopted. . .'.[40] All the applicable laws are to be administered and enforced by the Federal authorities, and the Federal district courts are given original jurisdiction of 'cases and controversies arising out of or in connection with any operations conducted on the outer Continental Shelf for the purpose of exploring for, developing, removing or transporting by pipeline the natural resources of the subsoil and sea bed. . .'.[41] In fact, many of the most difficult court decisions in this area have thus been those which have attempted to determine the extent to which Federal law, state law, or general maritime law is applicable.[42] Although it seems clear that the law to be applied is Federal law, supplemented by state law enforced as Federal law where there are gaps in the Federal law and the state law is applicable and not inconsistent, in some circumstances, again generally involving determination of whether or not the cause of action arose out of something connected with a ship or seaman, maritime law may be applied.[43]

Just as there are problems of overlap and conflict between Federal and state law, so there are overlaps and conflicts even between different federal statutes. A number of them which are relevant to employment are made expressly applicable in some cases by the Outer Continental Shelf Lands Act itself, but they often inter-relate in a curious manner. The OCSLA itself stipulates that:

> For the purposes of the National Labor Relations Act,[44] as amended, any unfair labor practice, as defined in such Act, occurring upon any artificial island or fixed structure. . .shall be deemed to have occurred within the judicial district of the adjacent state nearest the place of location of such island or structure.[45]

The Fair Labor Standards Act of 1938 also expressly provides for its application on the outer continental shelf.[46]

But the situation is not always so straightforward. Thus although the Occupational Safety and Health Act of 1970 provides that 'This Act shall apply with respect to employment performed in a workplace in. . . outer continental shelf lands. . .'.[47] It also stipulates that 'Nothing in this Act shall apply to working conditions of employees with respect to which other Federal agencies. . .exercise authority to prescribe or enforce standards or regulations affecting occupational safety or health.'[48]

The OCSLA in turn contains the express provision that:

> The head of the Department in which the Coast Guard is operating shall have authority to promulgate and enforce such reasonable regulations with respect to lights and other warning devices, safety

> equipment, and other matters relating to the promotion of safety of life and property on the islands and structures. . .as he may deem necessary.[49]

But no mention is made of provisions relating to health.

The OCSLA also provides that with respect to the disability or death of an employee resulting from any injury occurring as a result of operations on the outer continental shelf for the purposes of exploration etc. of the natural resources, compensation is payable under the provisions of the federal Longshoremen's and Harbor Workers' Compensation Act, but an employee is so defined as to exclude the master or member of the crew of any vessel or any employee of any state government or the Federal government or any of its agencies.[50] In relation to the same subject it should be noted that the Occupational Safety and Health Act stipulates that:

> Nothing in this Act shall be construed to supercede or in any manner affect any workmen's compensation law or to enlarge or diminish or affect in any other manner the common law or statutory rights, duties, or liabilities of employers and employees under any law with respect to injuries, diseases, or death of employees arising out of or in the course of employment.[51]

It should be noted, however, that no definition is provided in the OCSLA of 'artificial islands and fixed structure', and the definition of a 'deepwater port' in the Deepwater Port Act does not remedy the problem that a semi-submersible drilling rig might be neither an 'artificial island' or a 'fixed structure'. In practice, of course, it is unlikely that an oil company or drilling company would seek to evade the application of certain Federal laws on such a rig because of the powers of the Federal government through the licences. But the position of those rigs which are or might be classified as ships is not clear either, and employees on semi-submersibles[52] or on drilling ships might in some circumstances find it advantageous to be classified as 'seamen', as the Longshoremen's and Harbor Workers' Compensation Act[53] is less advantageous than the Jones Act,[54] and indeed there have been many cases of offshore workers seeking to bring themselves under its provisions.[55]

Thus, although the United States has long had a very thorough system for the regulation of activities on the continental shelf, it is fairly obviously not without its faults. It is inordinately complicated, beset by problems of definition, problems relating to jurisdiction, the overlapping of statutes, and the division of responsibilities for the administration and supervision of the activities between numerous different agencies with the Coast Guard, the US Geological Survey, the Corps of Engineers, and the Occupational Safety and Health Administration all having some responsibility for safety standards and their enforcement. As one team of researchers said when they examined the whole system of government management of the outer continental shelf,

> . . . . .the overall picture. . .is one of fragmentation. . . This results in a requirement for extensive inter- and intra-agency co-ordination to avoid gaps and oversight, particularly in matters concerning the relatively new concerns for the environment and safety, and places an enormous strain on the Department of the Interior if it attempts to be effective in meeting all of its responsibilities.[56]

2. The Countries Surrounding the North Sea

All the countries surrounding the North Sea have legislation of one kind or another relating to offshore exploration and production. Indeed, most have special rules extending the laws of the country in question to installations, and have forced upon the companies engaged in the activities, through licences or instruments made under statutory authority, regulations relating to the safety and employment of workers on the continental shelf. The main differences between the systems employed by these countries and that found in the United States is that most of the European countries have adopted a licensing or concession system for allocating rights to explore the continental shelf or produce from it, rather than a leasing system, and that nearly all the countries have a much less complicated system of administration and enforcement. It should be recalled that the sectors of the North Sea which are seeing the most activity, apart from that of the UK, are those of the Netherlands and Norway.

(a) The Netherlands. The Netherlands have mining legislation relating to onshore activities which dates back to the time of Napoleon which, although amended, is still in force as the basic law.[57] Separate legislation passed in 1965 governs offshore activities,[58] and the dividing line between their spheres of application is the limit of territorial waters.[59] Onshore the Dutch employ a concession system for exploration and production, although permits are needed and issued for drilling outside

areas held under concession,[60] which broadly speaking confer perpetual ownership of the mine on the concessionaire. As a matter of practice the Government has long used the concession deed to impose upon the concessionaire a number of obligations, including the observation of safety regulations, last published in 1964.

In relation to offshore activities the Continental Shelf Mining Law of 1965 provides for different kinds of licence for prospecting or producing. The Act lays down provisions relating to royalty payments, and also includes the provision that the licences may only be granted subject to regulations.[61] The Act further stipulates that regulations may be made by Order in Council covering nearly all aspects of offshore activities, including surveying,

(b) in the interests of the safety and the health of the persons employed in the survey or production, (and)

(c) concerning the work of the persons referred to under (b).[62]

The Act also provides for the establishment of safety zones around offshore installations,[63] and for an inspectorate with powers of entry and search,[64] enforced by Ministerial power to direct changes, improvements and rectifications in the activities.[65]

Detailed regulations pursuant to the Act were made in 1967.[66] They cover such matters as the delivery of the names of all supervisors, and the names and details of all employees, to the Inspectorate of Mines,[67] the posting of the regulations where they can be seen,[68] the power of the Minister to issue orders to licensees,[69] and the communication to the authorities of information regarding the operations.[70] Although sometimes expressed in general terms, the regulations impose requirements as to the repair of machinery,[71] the safety of persons on installations,[72] and the construction and equipment of the installations themselves.[73] They cover the making of the installations, and the lights they have to show,[74] the lighting of working places,[75] and the standards of machinery to be used.[76] Specific regulations on safety forbid the consumption of alcohol,[77] provide for the notification of accidents or near accidents,[78] and cover such matters as the provision of protective clothing.[79] There are special rules on diving operations,[80] on health and hygiene, sleeping arrangements, health care and so forth,[81] as well as on accidents and rescue services.[82] Regulations prescribe minimum ages for certain activities, and lay down specific work and rest hours.[83] All in all little is overlooked, but it should be noted that the Dutch also treat floating drilling rigs as vessels, and therefore subject to specific provisions covering ships.

In relation to the applicability of the general law to events on offshore installations, the Law for North Sea Installations of 1964 lays down that the Dutch penal code applies,[84] and that general provisions of the legal code can be extended or modified by administrative order.[85]

(b) Norway. Norway asserted sovereignty over the sea bed and subsoil of the submarine areas beyond the coast and up to the median line by Royal Decree on 31 May 1963. The country was even then under pressure to regularise the legal position so that offshore drilling operations could begin. There followed a series of temporary Acts and Decrees,[86] until the Royal Decree of the 9 April 1965 provided a complete legal framework for offshore operations, supplemented in 1967 by a Royal Decree covering safe working practices.

The great merit of the Norwegian system is that no distinction is drawn between operations within territorial waters, and those taking place beyond.[87] The 1972 Royal Decree,[88] which replaces the Royal Decree of 1965, lays down provisions which simply apply to all offshore operations. It basically provides for two kinds of licences, a non-exclusive reconnaissance licence which lasts for three years, and gives the holder the right to carry out surveys within a limited period of time but does not give drilling rights, and an exclusive production licence given in respect of blocks of the continental shelf for a period of six years. The Decree lays down provisions relating to fees, royalties and surrenders, and gives the Ministry of Industry and Handicrafts power to make further regulations which may relate *inter alia* to safety measures of any kind.[89] There is an inspectorate to secure compliance with the regulations, with power to order operations to be discontinued. Indeed, in the event of repeated non-compliance the Ministry is given the power to revoke the licence.[90] The Decree also establishes safety zones around all installations on the Norwegian continental shelf.[91]

In relation to the application of the law to activities on the shelf, the Norwegians have always been straightforward. The 1972 Decree thus lays down that unless otherwise provided Norwegian law applies to all installations and appliances on the continental shelf, to activities carried out on them, and within the safety zones.[92] Furthermore, in the event of 'damage or inconvenience' being caused, the Norwegian law of torts is expressly made applicable.[93]

The basic safety regulations are contained in the Royal Decree of 25 August 1967, and are just as extensive as the Dutch equivalents, but are again marred by numerous references to 'good oil-field practice' and other such generalisations.[94] Nevertheless they do cover most aspects of work on drilling platforms, defined as 'any installation, including vessels, equipped for drilling for petroleum'.[95] Thus the Decree covers safety equipment,[96] living quarters and working spaces,[97] markings and lights,[98] requires the keeping of records of the personnel on the platform,[99] the presence of a standby vessel,[100] and covers divers and diving activities.[101] There is even a provision requiring the oncoming shift to be informed of safety problems by the shift going off.[102]

The Decree also gives authority to the Ministry of Industry to make other regulations where these might be necessary.[103] The Ministry is in fact broken down into a number of Directorates, and these and other government bodies have been allocated differing responsibilities under this and other Decrees. Thus the Maritime Directorate has issued detailed regulations covering the construction and design of rigs, including lifesaving equipment. The Petroleum Directorate issues regulations covering drilling equipment, and the Directorate of Labour Inspection issues regulations relating to cranes, boilers, personal protective equipment, organised safety precautions and diving operations. The Directorate of Health covers hygiene conditions and first-aid and hospital equipment. Although the result is extensive and detailed regulation of all offshore operations, with minute specifications in many instances, the overall result requires a good deal of inter-agency co-operation and administration that in some instances parallels the American situation.

(c) Denmark. The exploration and exploitation of the Danish part of the continental shelf is now governed by the provisions of the Royal Decree of the 7 June 1953 on the Exercise of Danish Sovereign Rights over the Continental Shelf, as amended by legislation passed in 1971.[104] Under the terms of the Decree as originally promulgated the provisions of legislation passed in 1950[105] were extended to the continental shelf allowing for the grant of exclusive concessions for up to fifty years at a time subject to various royalties and payments. Under these provisions the single concession granted in 1962 to a consortium headed by A.P. Møller in respect of the entire area of Denmark was extended to the continental shelf as well.[106]

Under the law of 1971 the concession, which was granted for fifty years, continues in force in accordance with the provisions of the Act of 1950.[107] Under the concession agreement itself the permission of the Minister of Transport has to be obtained before any installation is established on the shelf,[108] and it provides that supervisors from the

Ministry may oversee the activities.[109] But the law of 1971 appears to supercede this essentially private relationship, and provides that the Minister of Public Works may make regulations concerning safety matters in connection with the setting up and operation of installations,[110] and in relation to the establishment of safety zones around them.[111] The law also stipulates that Danish law shall apply to installations used for exploration and exploitation of the continental shelf and in the safety zones.[112] Since the law was passed there has in fact been less and less activity in the Danish sector, and that country's interests have shifted to the continental shelf off Greenland and the Faeroe Islands.

(d) France. After the French adhered to the Geneva Convention in 1965[112a] special legislation was enacted in 1968 dealing with offshore exploration and production, and laying down special rules for offshore installations.[113] Basically this law extends the provisions of the Mining Code to offshore operations, but makes a number of alterations for such activities relating to such topics as customs duties,[113a] royalty payments,[114] and the duration of concessions and exploration permits, and the process of applying for them.[115]

The special regime laid down by the law of 1968 applies to all platforms and other installations involved in exploration or exploitation, and to ships participating directly in such operations.[116] All platforms are subject to the laws relating to the safety of life at sea, and if they float, are otherwise subject to the laws relating to registration and sailing permits as well as laws preventing boarding whilst afloat.[117] For these purposes the person in charge is designated the captain,[118] and he and the owner or user of the installation are all made responsible for the installation, its functioning, and the maintenance of its signals,[119] and must conform to instructions from competent authorities in these respects.[120] Provision is also made for the establishment of safety zones around installations, and for the control of activities within them.[121]

Under the law of 1968, French law and regulations apply to all installations and equipment, as well as to the security zones around them, as though they were located on the territory of metropolitan France itself.[122] Further regulations specify that all those persons who exercise responsibilities for the enforcement of laws and regulations that might therefore be applicable to offshore activities are to exercise their functions on the continental shelf in the same way.[123] The law itself specifically applies the provisions of criminal law, and criminal procedure, in force at the place where the courts exercising jurisdiction over the port to which an installation is attached are held,[124] and the regulations indicate further that the powers of the courts covering the point of the coast nearest to an installation are extended out to the installation itself.[125] Specific provision is also made for workers on board offshore installations and ancillary equipment to be brought under the provisions of the social security system relating to seamen if they so wish and to take advantage of certain provisions of the Maritime Labour Code relating to illness, injury and retirement.[126] It should be noted that the law also provides for further regulations to be issued relating to the application of law on the installations, and for the conduct of operations.[127]

(e) Belgium and Germany. Although there is at present little activity in either the Belgian or German sectors of the North Sea, both these countries have enacted legislation covering offshore activities. The Belgian law[128] came into effect in June 1969, exerting sovereignty over the continental shelf, and providing that it may be allocated by concessions given by the King under such terms and conditions as he thinks fit. It also provides that installations and other equipment located on the continental shelf, and the persons on them, are subject to Belgian law, the criminal law is specifically extended, and for non-criminal purposes acts or events are deemed to have taken place in Belgium. It also provides for the creation of safety zones around installations. The German law, which has now been amended several times,[129] provides for a system of permits and for the enforcement of various terms and conditions, and for a system of administrative orders which may be issued in respect of the activities carried out under the law.

(f) The Republic of Ireland. The principal legislation dealing with petroleum exploration and production in the Republic of Ireland is the Petroleum and Other Minerals Development Act 1960. By this Act all petroleum is vested in the Minister for Industry and Commerce,[130] and no one but the Minister is empowered to search for petroleum without holding an exploration licence, a petroleum prospecting licence or a reserved area licence, or being the lessee under a petroleum lease which alone gives the right to produce.[131] All these licences may be issued by the Minister 'upon such terms and conditions. . .as (he) thinks fit',[132] and the Act contains provisions relating to rents and royalties.[133] From the practical point of view those provisions relating to the compensation to be paid in respect of private property rights in petroleum, and those sections relating to the compulsory acquisition of

land are obviously of considerable import.[134] It should be noted, too, that the Act empowers the Minister to make regulations relating to safety measures and such other matters in relation to petroleum operations as he thinks necessary.[135] The Minister also has powers of direction by notice where operations are being conducted in such a manner as to cause, or be likely to cause, unnecessary damage.[136]

The provisions of this Act are extended to the continental shelf by the Continental Shelf Act 1968,[137] which in many respects is very close to the legislation of the United Kingdom. It vests rights in the sea bed and subsoil in the Minister,[138] to be exercised in designated areas of the continental shelf.[139] Under the Act the Minister of Transport and Power must give consent for the construction, alteration or improvement of any structure or works in designated areas, and for the removal from such an area of any object or material.[140] Provision is made for enquiries and for refusal of permission where the action in question would cause obstruction or be a danger to navigation,[141] and as is the case with the legislation of most of the countries around the North Sea, specific provision is made for the establishment of safety zones around installations.[142]

Much of the Act is in fact taken up with the application of other laws to operations on the continental shelf. Thus the Act specifically extends the social welfare legislation of the Republic, including the Insurance (Intermittent Unemployment) Acts.[143] But in relation to the general law the Act unhappily echoes the UK legislation,[144] providing that any 'act or omission' taking place on an installation in a designated area or any waters within five hundred metre of one, or on any waters outside the five hundred metre zone or under or above any waters or installations in a designated area 'in connection with the exploration of the sea bed or subsoil or the exploitation of their natural resources', is deemed to take place in the Republic, if it would constitute a wrong[145] or a criminal offence if it had taken place there, for all purposes related to the wrong or the offence.[146] Proceedings for offences may be brought in any part of the State,[147] and it is further provided that: 'Any jurisdiction conferred on any court under (section 3) shall be without prejudice to any jurisdiction exercisable apart from (the) section by that or any other court.'[148]

As is the case with the legislation of the United Kingdom, these provisions mean that in many cases the application of other Acts and regulations relating to labour and safety matters is dependent on the inclusion in those instruments of specific reference to continental shelf activities. That in turn requires legislative initiative that is not always forthcoming.

C. The Legal Framework for the UK Sector

The legislative framework for offshore exploration and oil and gas production in the UK sector of the continental shelf is similar in many respects to that of other countries around the North Sea. It provides for the extension of the legal provisions initially designed for the petroleum industry onshore, but has been amended to take account of the rapid development of the offshore industry and government policies for participation and regulation. Separate legislation relating to safety and working conditions on offshore installations has also been passed, and the two aspects can conveniently be discussed separately.

1. The Continental Shelf Act 1964

The basic legal framework for offshore activities is provided by the Continental Shelf Act 1964, which vests in Her Majesty '. . .any rights exercisable by the United Kingdom outside territorial waters with respect to the sea bed and subsoil and the natural resources, except in so far as they are exercisable in relation to coal. . .'.[149] According to section 1 (7): 'Her Majesty may from time to time by Order in Council designate any area as an area within which the rights mentioned in subsection (1) of this section are exercisable, and any area so designated is in this Act referred to as a designated area.'[150]

These provisions have enabled the Government to create a system of licences for the offshore exploration and production of petroleum and natural gas both inside and outside territorial waters[151] within the median lines drawn between Britain and the other countries surrounding the North Sea. Although there are differences in the system as it applies onshore and offshore, this is achieved through the extension of the Petroleum (Production) Act 1934.[152] Thus section 1 (3) of the Continental Shelf Act provides: ' In relation to any petroleum[153] with respect to which those rights[154] are exercisable sections 2 and 6 of the Petroleum (Production) Act 1934 (which relate to the granting of licences to search and bore for, and get, petroleum) shall apply as they apply in relation to petroleum in Great Britain.' Section 2 (1) of the Petroleum (Production) Act gives the Board of Trade (now the Department of Energy)[155] power to grant licences 'to such persons as they think fit', and by section 2 (2) 'for such consideration. . *.and upon such other terms and conditions as (they) think fit'.*[156] Section 6 instructs the Department, before granting any licences under the Act, to make regulations as to the manner of application for a licence, the conditions as to the size and shape of the areas, and more important, to draw up 'model clauses which shall, unless the (Department) think fit to modify or exclude them in a particular case, be incorporated in any such licence.'[157] It should be noted that by section 1 (4) the 'Model clauses under section 6 of the Petroleum (Production) Act 1934 as applied by the preceding subsection shall include provision for the safety, health and welfare of persons employed on operations undertaken under the authority of any licence granted under that Act as so applied.'

The effect of these provisions is apparent in the Petroleum (Production) Regulations 1966[158] (as amended)[159] which fit them together. One of the most important requirements made by these regulations is that the licence holder must be a UK citizen or a company incorporated in the UK.[160] This reflects the difficulty that the Department of Trade and Industry thought it would have in exercising control over the continental shelf operations of foreign companies and personnel. It was felt essential to have at least one person or entity the Department could hold liable should that be necessary.[161] It must be borne in mind that though most of the companies that operate in the North Sea have a great deal of expertise, little of this was found in the United Kingdom, at least initially. The result was that those companies that were not UK international companies either brought crews from their operations in other parts of the world, or sub-contracted the drilling work out to firms who specialise in this type of work, and many of the operators are therefore either Dutch, French, or (predominantly) American.

The schedules to the regulations set out the 'model clauses' for each of the different kinds of licences, i.e. the form which each one is to take,[162] and it is here that the Government's policy on royalty payments has been brought into effect through amendments included in the Petroleum and Submarine Pipe-lines Act 1975.[163] In fact, of course, the licences include provisions relating to all aspects of the operations, and cover such matters as development and production programmes, the commencement and abandonment of wells, records, returns and samples, the appointment of a responsible operator for the licensee,[164] and other working obligations. Of particular relevance here is the clause found in each one relating to the safety, health and welfare of the employees,[165] which is as follows: 'The Licensee shall comply with any instructions from time to time given by the Minister in writing for securing the safety, health and welfare of persons employed in or about the licenced areas.'[166]

The various instructions given by the Minister are examined below with the other provisions relating to safety, health and welfare. But it is important to note, especially as the Mineral Workings (Offshore Installations) Act 1971 was passed to replace these instructions with proper regulations, that there are only two remedies available to the Government until such time as such regulations come into force.[167] All the licences contain clauses giving the Minister the right in the event of non-compliance with instructions to either '. . .execute any works and to provide and install any equipment which in the opinion of the Minister may be necessary to secure performance of the. . .obligations. . . and to recover the expenses of so doing from the licensee'[168] or revoke the licence.[169] Thus the paramount defect of the system is that the instructions are part of a contract between the Government and the licensee, and the only remedies available are those of a contractual nature. In fact entry and rectification is unlikely because of the difficulties of operations against the background of commercial pressure (part of which is created by the Government itself), and the relative lack of experience of the inspectorate and the limited extent of its surveillance. In much the same way revocation is a sledgehammer remedy that would be unlikely except in the most extreme of cases where the benefits to the Government of continuation do not outweigh those of stopping the operations in a situation where at present the Government is almost entirely reliant on the large oil companies for the development of the industry. There remains, too, the serious defect that such contractual remedies confer no rights on third parties, such as employees. Even though the regulations intended to benefit those same third parties,[170] they could not sue under the licences.[171]

2. The Mineral Workings (Offshore Installations) Act 1971

In relation to employment on the continental shelf, and to activities on offshore installations themselves, the principal legislation is the Mineral Workings (Offshore Installations) Act 1971, which will be discussed throughout the remainder of this work. It is in fact primarily a piece of enabling legislation, designed to provide maximum flexibility, and to take effect through regulations[172] which will eventually put an end to the uncertainties that at present surround the safety provisions in the licences under the continental shelf and petroleum legislation.

The preamble describes the Act as one intended: '. . .to provide for the safety, health and welfare of persons on installations concerned with the underwater exploitation and exploration of mineral resources in the waters in or surrounding the United Kingdom, and generally for the safety of such installations and the prevention of accidents in or near them'.

The impetus for the legislation came from the North Sea itself. The high number of accidents to rigs, and the hazards to which even oil men with experience elsewhere have had to learn to adapt, provided a spur to action.[173] It is indeed far from surprising that one member of the House of Commons said that the Bill had been coming since 1961,[174] despite the fact that actual drilling did not start until 1963. In fact, the greatest encouragement came from the foundering of the drilling rig 'Sea Gem' in December 1965. A Tribunal of Inquiry set up to examine the causes of the accident recommended that there should be a statutory code of safety for offshore rigs supported by credible sanctions,[175] and subsequent amendment of the licences could not be expected to meet all the points made in the Report. In the Queen's Speech of 1969, the Labour Government announced its intention to bring in legislation, but failed to find room for the Bill before it went out of office. The incoming Conservative administration was able to give it high priority because of the absence of opposition, and such criticism as there was, mainly from John Prescott and Kevin MacNamara (Labour MPs for Hull) was almost entirely directed at the adequacy of various safety measures, and at the disciplinary provisions.[176]

The Act is intended to apply to both inland waters as well as offshore operations,[177] but only to the exploration and exploitation of mineral resources.[178] It extends only to ' any floating or other installation which is maintained in the water, or on the foreshore or other land intermittently covered with water, and is not connected with dry land by a permanent structure. . .providing access at all times. . .',[179] i.e. truly 'offshore' installations which are 'maintained, or. . .intended to be established, for underwater exploitation or exploration'.[180] Indeed, the definition provided by the Act is drawn in terms sufficiently wide to cover all the varieties of platforms from which drilling out at sea can possibly take place. It is worth bearing in mind that there are four broad types. Firstly, there is the 'jack-up' type – a platform carrying the drilling rig which is floated to the drilling position, the legs are wound down and the platform is then elevated to a safe distance above the water. Secondly, there is the 'semi-submersible' type, generally speaking a floating pontoon large enough to remain stable in almost any weather, and which is anchored to the bottom on all sides. Thirdly, there are fixed production platforms which are built on to the sea bed and can act as terminals for the pipes to the shore as well as providing a platform for further development drilling. These have legs which are concreted into the sea bed, and are intended as permanent, immovable structures. Exploration is also carried out from ships. These will normally be registered as ships under the Merchant Shipping Act 1894, but are covered by the Act for the sake of completeness, though only when they are anchored to the sea bed and are engaged in actual drilling. They pose far less of a problem in regard to law and safety: the law of the flag is a great deal less complicated in application than the Continental Shelf Act, and ships have their own Code of Safe Working Practices. In practice, the specialised Code, and the regulations under Mineral Workings (Offshore Installations) Act are applied when the drill ship is actually drilling and the Merchant Shipping Acts at other times. This means, unfortunately, that a choice of legislation may have to be made at some time, though this is provided for in the Mineral Workings (Offshore Installations) Act.[181]

Other sorts of 'installation' which are intended to be covered include suction dredgers. This was made clear during the passage of the Bill through the House of Lords. Lord Ferrers said: 'In so far as the dredger falls within that category whereby it is not connected to the land and is in fact dredging for minerals, then the dredger would come within the purview of this (Act).'[182]

That the platform is referred to as an 'installation' might be taken to mean that it has to be installed. Indeed, section 8 (5), which seeks to extend the application of that section to instances when an 'installation' is in transit, would seem to bear out this interpretation. There is a conflict in this case with section 1 (3) (b) which extends the meaning of 'offshore installation' to an installation 'intended to be established', but would thus seem to exclude an installation being removed.[183] Certainly the Government has an interest in platforms when they are being moved from one well to another as they always have men aboard during the tow.[184] This is especially so if they are not registerable as ships. Furthermore, if this interpretation is incorrect the question arises of when an installation is 'installed'. No definition is provided by the Act, but it could be either when it puts down its anchor, when it is ready to drill, or indeed when drilling has commenced. It is submitted that it is preferable to assume that the Act applies to installations on the move as well as those 'spudded-in', i.e. actually drilling a well.

In fact, it would appear that the definition of an installation is also intimately bound up with whether or not it is possible to have people aboard it. Section 1 (1) (b), when it speaks of an installation being 'maintained', and section 12 (2), which further extends the definition of an offshore installation to include 'any part of an offshore installation capable of being manned by one or more persons', would seem clear evidence of the fact that the Government is mainly interested in installations when there are people on them,[185] and indeed section 12 (3) (as amended by section 44 (3) of the Petroleum and Submarine Pipe-lines Act 1975) now takes structures which are essentially only well-heads and incapable of being manned, outside the definition. It provides that references in the Act to an offshore installation

> . . .do not include an installation which at the relevant time consists exclusively of a pipe-line, whether or not any part of it previously formed part of an offshore installation, and do not include any apparatus and works which are treated as associated with a pipe or system of pipes in pursuance of section 33 (1) of the Petroleum and Submarine Pipe-lines Act 1975 and are at the relevant time incapable of being manned by one or more persons.'

There is a clear distinction between 'maintained' and 'manned', and indeed some production platforms are unmanned, but visited at regular intervals for servicing and checking. Nevertheless they are 'maintained' as people do work on them from time to time, and are 'capable of being manned'.[186]

3. Submarine Pipe-lines and Pipe-laying Activities

Under the Petroleum and Submarine Pipe-lines Act 1975, the provisions of the Mineral Workings (Offshore Installations) Act have been extended, and a new regime created for pipe-lines and pipe-laying in 'controlled waters', i.e. territorial waters and any designated area under the Continental Shelf Act.[187] The construction and use of pipe-lines begun after the Act came into force require the written authorisation of the Secretary of State in response to an application made in the proper fashion.[188]

But the Act not only gives the Secretary of State broad control over all activities relating to pipe-lines (for example, he may order modifications to be made to a pipe-line to increase its capacity,[189] and give to a person other than the owner of a pipe-line the right to use that pipe-line[190]), but also confers upon him wide powers as regards the terms of the authorisations,[191] and the contents of regulations which he may make '. . .for the purpose of securing the proper construction and safe operation of pipe-lines, preventing damage to pipe-lines and securing the safety, health and welfare of persons engaged on pipe-line works. . .'.[192]

These powers, at least in their relation to offshore workers, are more appropriately discussed later in this work.[193] But the scope of the legislation should be noted. The definitions provided by the Act are very full. Thus a pipe-line means (except where the context otherwise requires) '. . .a pipe or system of pipes (excluding a drain or sewer) for the conveyance of any thing, together with any apparatus and works associated with such pipe or system. . .'.[194] This definition is expanded by the provision that:

> . . .the following apparatus and no other shall be treated as associated with such a pipe or system, namely –
>
> (a) Any apparatus for inducing or facilitating the flow of any thing through, or through a part of, the pipe or system;
> (b) valves, valve chambers and similar works which are annexed to, or incorporated in the course of, the pipe or system;
> (c) apparatus for supplying energy for the operation of any such apparatus or works as are mentioned in the preceding paragraphs;
> (d) apparatus for the transmission of information for the operation of the pipe or system;
> (e) apparatus for the cathodic protection of the pipe or system; and
> (f) a structure used or to be used solely for the support of a part of the pipe or system.[195]

The Secretary of State is also given power to provide that a specified part of a pipe-line is to be treated as a pipe-line,[196] to modify the application of the Act to pipe-lines of a prescribed kind or until specified conditions are met,[197] and indeed to extend the provisions of the Act, with or without modifications, to pipe-lines or parts of pipe-lines which do not start or begin in the UK or controlled waters.[198] Such an extension can only contain provisions consistent with the jurisdiction which belongs to the UK under international law.[199]

In relation to this work, however, it is the definition of 'pipe-line works' which is of more importance. According to section 26 (2) this means

> (a) assembling or placing a pipe-line or length of pipe-line;
> (b) inspecting, testing, maintaining, adjusting, repairing, altering

or renewing a pipe-line or length of pipe-line;

(c) changing the position of or dismantling or removing a pipe-line or length of pipe-line;

(d) opening the bed of the sea for the purposes of works mentioned in the preceding paragraphs, tunnelling or boring for those purposes and other works needed for or incidental to those purposes;

(e) works for the purpose of determining whether a place is suitable as part of the site of a proposed pipe-line, and the carrying out of surveying operations for the purpose of settling the route of a proposed pipe-line.

Although the regulations under section 26 relating to pipe-line works '. . .may include provision with respect to the use of any aircraft, vessel, vehicle, structure, plant, equipment or other thing for the purposes of any pipe-line works and with respect to the movement of and the precautions to be taken on or in connection with any of those things. . .'[200] the section also specifies that

> References to a pipe-line in the preceding provisions of this section are to a controlled pipe-line but shall be construed as excluding –
>
> (a) any pipe-line so far as it forms part of –
>
> (i) an offshore installation within the meaning of the Mineral Workings (Offshore Installations) Act 1971, or
> (ii) such other installation as is mentioned in section 44 (1) of this Act, or
> (iii) the equipment of a vessel or vehicle; and
>
> (b) any apparatus and works associated with a pipe or system of pipes and prescribed for the purposes of this paragraph.[201]

This subsection indicates clearly that the intention of the legislation was to separate administration of the pipe-lines themselves and the apparatus used to place them on the sea bottom, from the production platforms which might be connected to them. In practice, of course, there would appear to be some overlap, and the point at which a pipe-line becomes part of an offshore installation is somewhat hazy. Indeed, section 44 (1) of the Act provides that:

> The Mineral Workings (Offshore Installations) Act 1971, except section 1, shall have effect as if any reference to an offshore installation within the meaning of that Act included reference to any other installation, whether floating or not, which –
>
> (a) is maintained or intended to be established, in controlled waters or waters in the United Kingdom, for use in connection with the conveyance of things by means of a pipe constructed in or under the sea; and
>
> (b) is, or when established will be, capable of being manned by one or more persons;. . .

This appears to make it clear that not only are platforms used for development drilling (as opposed to exploratory drilling) considered offshore installations, but so are platforms and other intermediate production structures such as buoyant towers which represent a production link to the underwater field half way between a subsea well-head (which would now be considered as a pipe-line) and a full platform. These structures most frequently consist only of a single leg tower with a heavy base secured to the sea bed and a small valve deck beneath a helicopter landing pad. Nevertheless the most important point of the legislation in this respect is the extension of regulatory powers to pipe-laying barges which fell outside the definition of offshore installations and were either not registerable as ships, and hence subject to the Merchant Shipping Acts, or for which regulations made under those Acts would be inappropriate.

II. THE OPERATION OF BRITISH LAW ON THE CONTINENTAL SHELF

A. Extraterritorial Application of Law

The question of the extraterritorial application of the law of the United Kingdom is fundamental to the determination of its effect on those who work or operate on the continental shelf outside territorial waters. It is, moreover, a complicated problem, but one towards which the Continental Shelf Act 1964 has gone some way in providing a solution.[202]

As far as legislation is concerned, the general rule appears to be as stated in *Maxwell*:

> In the absence of an intention clearly expressed or to be inferred

> either from its language or from the object of the subject matter or history of the enactment, the presumption is that Parliament does not design its statutes to operate on its subjects beyond the territorial limits of the United Kingdom.[203]

However, this is only a presumption, and important though it may be, it remains rebuttable. Indeed there are a number of instances in which statutes relating to civil matters have been held to apply extraterritorially.[204] These relate mostly to problems of family law, and in the cases concerning labour legislation the proposition that the law should apply outside the United Kingdom failed to gain acceptance.[205]

Similar considerations apply to the criminal law. Viscount Simonds said in *Cox* v. *The Army Council:*[206] 'Apart from those exceptional cases in which specific provision is made in regard to acts committed abroad, the whole body of the criminal law of England deals only with Acts committed in England.' As a correct statement of the law it is perhaps slightly misleading. There are indeed several exceptions, including for example, murder and manslaughter.[207]

As far as procedure is concerned the Rules of the Supreme Court place a severe limitation on the possibility of service of a writ out of the jurisdiction,[208] though it will be permitted[209] if relief is sought against a person domiciled or ordinarily resident within the jurisdiction of the court.[210] It is this which it is necessary to determine, for the jurisdiction is normally coincident with the territorial limits. In fact, even if the courts will entertain the writ at common law they may have no power to grant a remedy.

This study is not concerned, however, with acts, criminal or otherwise, which take place abroad in the sense that they occur within the jurisdiction of foreign courts. The problem of the application of laws to the continental shelf is that the British sector is clearly extraterritorial in so far as it is, by definition, beyond the limits of territorial waters. Nevertheless, the UK continental shelf is an area in which the jurisdiction of the United Kingdom over offshore installations has been acknowledged by an international convention. It might therefore be aptly described as a sort of legal 'no man's land' for which the coastal state has been given authority to make certain limited legal provisions. This has now been exercised in section 3 of the Continental Shelf Act.

The Act attempts to meet some of the points made here in two different ways. Firstly, in section 3 (1) it applies the full range of the criminal law to events taking place on an offshore oil-rig by means of a legal fiction by which anything done on an installation which would be a criminal offence if done ashore is to be treated as though it took place ashore. The subsection reads:

> Any act or omission which –
>
> (a) takes place on, under or above an installation in a designated area or any waters within five hundred metres of such an installation; and
> (b) would, if taking place in any part of the United Kingdom, constitute an offence under the law in force in that part, shall be treated for the purposes of that law as taking place in that part.

It is reinforced by section 11 (1) which states that:

> Proceedings for any offence under this Act (including an offence under another Act as applied by or under this Act and anything that is an offence by virtue of section 3 (1) of this Act) may be taken, and the offence for all incidental purposes be treated as having been committed, in any place in the United Kingdom.

Similar language is employed in section 29 (1) of the Petroleum and Submarine Pipe-lines Act 1975 in relation to offences committed in relation to pipe-line works by virtue of regulations made under that Act.

The use of this 'deemed place' fiction[211] seems to preserve the rule relating to extraterritorial effect rather than to break it. It means that an act of stealing, for example, which occurs on an installation may be prosecuted in any convenient court on shore,[212] and be deemed for the purposes of bringing the proceedings to have occurred in that place and not out at sea. Clearly there is a distinction between applying the criminal law directly, and using this 'deemed place' fiction. But for practical purposes in a magistrates court the use of the fiction will make no difference. The Act enables the court to try the case, and if the defendant is before it, no problems as to sentence and enforcement should arise. For this reason, indeed, section 11 (3) of the Act gives any police constable the same powers, protection and privileges on an installation as he has in the area for which he is a member of the police force.

In relation to the civil law, the Act proceeds in a rather different fashion. Section 3 (2) attempts, in the words of the sidenote, to apply the civil law to 'acts and omissions' which take place on installations in a designated area.[213] The subsection reads:

> Her Majesty may by Order in Council make provision for the determination in accordance with the law in such part of the United Kingdom as may be specified in the Order, of questions arising out of acts and omissions taking place in a designated area or in any part of such an area in connection with the exploration of the sea bed or subsoil or the exploitation of their natural resources, and for conferring jurisdiction with respect to such questions on courts in any part of the United Kingdom so specified.

It is not immediately apparent just what the section has achieved, but some answer is provided by the Order made under this section.[214] This states that: 'Subject to section 3 (1) of the Act, the law in force in England shall apply for the determination of questions to which section 3 (2) refers arising out of acts and omissions taking place in the English area. . .'.[215]

This wording might seem to imply that the civil law is applied directly to events on installations. But in fact it applies the law 'for the determination of questions. . .'. The practical differences would not seem to be very great. Even if the law is not extended in so many words, the fact that it can be applied has almost the same result. This is reinforced by the subsequent paragraph which stipulates that: 'The High Court shall have jurisdiction for the determination of any questions to which section 3 (2) refers and which. . .fall to be determined in accordance with the law in force in England, as it would have if the acts or omissions in question had taken place in England. . .'.[216]

This is not the same use of a 'deemed place' fiction. But jurisdiction of the court is extended by this wording, and, effectively, so is the law which it can therefore apply. This has important repercussions if certain other limitations imposed by the language of the section cannot be overcome.

Firstly, by confining the application of the subsection to 'acts and omissions' the clarity of the similar formula used in section 3 (1) is lost. Not all events would seem to be brought under the jurisdiction of the courts. For example, though a contract made in England can be broken anywhere in the world, giving the right to sue in the English courts,[217] does the law of England apply to a contract made on an offshore installation? Is the formation, or the breach of such a contract, an 'act or omission'? This would need further judicial clarification,[218] although the problem is entirely academic in that the only question a court would raise faced with a contract made on an offshore installation would be 'which law applies to the contract?'. This can be solved without taking exclusive notice of the place it was made.

Secondly, the 'acts and omissions' to which the regulations refer are limited in the Act itself to those which 'take place. . .in connection with the exploration of the sea bed or subsoil or the exploitation of their natural resources. . .'. It is the phrase 'in connection with' which gives rise to particular difficulties. In the Crown Proceedings Act 1947, for example, the phrase 'connected with' is used in relation to the duties of members of the armed forces[219] and bears the meaning of 'in the course of'. By contrast, the phrase 'connected with' is used in a different way in the Trades Disputes Act 1906 where it means 'concerning directly'.[220] Clearly then, the phrase 'in connection with the exploration. . .etc.' is capable of either a broad or a narrow interpretation by the courts.

In using this phrase the intention of the legislature may well have been only to exclude liability in situations where the worker was 'off on a frolic of his own' in a situation where the worker cannot leave the place of employment.[221] But whereas under the order one would think that the result is to give a right of action for damages to a worker injured on an installation through the negligence of a contractor, through a defect in a piece of equipment, through the collapse of the structure of the installation, or through the negligent action of one of his workmates, this might not be the case. One has first to answer such questions as 'is an offshore installation engaged in exploration or exploitation of the sea bed or subsoil only when it is drilling?' A restrictive answer might deny a worker a right to action in tort[222] when the rig was not actually drilling, but moving from one well to another possible site.

In the face of such ambiguity it is submitted that the broadest possible interpretation of the phrase is the best. It would be absurd to make legal rights dependent upon the drilling programme of an installation, and quite clearly the presence of the rig on the continental shelf is unlikely to be for anything other than the exploration or exploitation of the sea bed and its resources. Exceptions to this general rule would appear to be those rigs involved in sinking airshafts for a Channel tunnel, for example, although those used to take rock samples might be covered by the term, 'exploration'.

There are three further points of interest. Firstly it should be noted that the waters to which the Mineral Workings (Offshore Installations) Act applies include those inside territorial limits as well as those in designated areas under the Continental Shelf Act, and in relation to section 3 of the 1964 Act there are several specific provisions extending its coverage. Section 8 (1) of the 1971 Act provides that section 3 of the Continental Shelf Act 'shall have effect as if any references in that section to an installation in a designated area included references to any offshore installation either in territorial waters of the United Kingdom or in a designated area.'

Furthermore, in relation to questions arising out of acts or omissions taking place on, under or above an offshore installation in territorial waters, or territorial waters within five hundred metres of an offshore installation, section 3 (2) of the 1964 Act also applies by virtue of section 8 (2) of the 1971 Act. No essential difference is therefore drawn in relation to either the Mineral Workings (Offshore Installations) Act or section 3 of the Continental Shelf Act between territorial waters and waters covering the continental shelf beyond those limits.

Secondly, it should be noted that the Continental Shelf Act reaches out to a potentially broader range of offshore equipment than the Mineral Workings (Offshore Installations) Act. Section 3 (1) of the 1964 Act applies to 'installations' (and indeed, by virtue of section 8 (5) of the 1971 Act, to installations in transit). No definition is provided in the Act of an 'installation', but the language of section 8 (1) of the 1971 Act (above) would appear to indicate that it is broader than that of an 'offshore installation' under the 1971 Act, even though section 44 (1) of the Petroleum and Submarine Pipe-lines Act extends the definition of the latter. It might therefore cover a structure which would not normally be thought to be covered as regulations made under section 44 (4) of the Petroleum and Submarine Pipe-lines Act had determined that it was incapable of being manned even though people had managed to get on it. Section 3 (2), of course, is not limited in its application to installations of any kind, but only to 'acts or omissions etc.'

The most important question raised by section 3 of the Continental Shelf Act, however, is related to the extent to which section 3 (2) operates to rebut the general legal presumption against the extraterritorial effect of statutes, resulting in an implied extension of other laws to rigs on the continental shelf. As the courts apply the law in force in England for the determination of questions arising out of acts and omissions in connection with exploration and exploitation of the sea bed and its resources, the answer must be that those laws which are relevant to the acts and omissions must be applicable. Thus statutes such as the Law Reform (Contributory Negligence) Act 1948 are as applicable to events on board an installation as the common law of torts. If this is the case then it will be for the plaintiff in any action to show that not only is the law relevant to acts or omissions, but in some cases also that a statute can be interpreted to cover situations on an offshore installation.

Thus, for example, there would appear to be no justification for the application of the Factories Act 1961 as an installation would not appear to come within the definition of a 'factory' given in the Act,[223] although the statute covers certain acts and omissions. On the other hand it is submitted that an installation does fall within the definition of 'premises' or 'fixed or movable structures' given in the Occupiers Liability Act 1957.[224] Further support for this approach is to be found in section 11 (3) of the Mineral Workings (Offshore Installations) Act which expressly provides for actions for breach of any statutory duty imposed by the Act causing personal injury, 'without prejudice to any action which lies apart from the provisions of this Act'.[225] However, these statutes all have the common feature of being specifically concerned with acts and omissions falling clearly within the province of the law of tort. Other legislation in the province of labour law merits further consideration.

B. Application of the Law to Foreigners Working on the Continental Shelf

In the present industrial context, where many of the operating companies are foreign (and especially American) owned, a large number of foreigners work on offshore installations in the British sector of the continental shelf. It is therefore important to determine to what extent British law applies to them.

In fact it is difficult to generalise about the application of the law to foreigners even in England.[226] In the case of statutes confering rights, foreigners have been allowed to take advantage of them in some cases,[227] and not in others.[228] In the case of statutes imposing liability, foreigners within the jurisdiction of the courts receive no special treatment,[229] but there is a presumption that a statute does not apply to aliens outside the jurisdiction. As James L.J. said:

> It appears to me that the whole question is governed by the broad, general, universal principle that English legislation, unless the

> contrary is expressly enacted or so plainly implied as to make it the duty of an English court to give effect to an English statute, is applicable only to English subjects *or to foreigners who by coming into this country. . .have made themselves during that time subject to English jurisdiction. Every foreigner who comes into this country. . .is. . .*entitled to the protection of the Sovereign and *subject to all the laws of the Sovereign.* But, if a foreigner remains abroad. . .it seems to me impossible to imagine that the English Legislature could have ever intended to make such a man subject to particular legislation.[230]

The problem is, therefore, bound up once again with the question of the jurisdiction of the courts.

In relation to the civil law, as section 3 of the Continental Shelf Act and the regulations made under that section confer jurisdiction on the courts to determine questions arising out of acts and omissions taking place on an offshore installation connected with the exploration and exploitation of the sea bed, the law which is made applicable by this method also applies to foreigners. This is reinforced by section 8 (3) of the Mineral Workings (Offshore Installations) Act, which states that:

> It is hereby declared that, notwithstanding that the said section 3 may affect individuals or bodies corporate outside the United Kingdom, it applies to any individual whether or not he is a British subject, and to any body corporate whether or not incorporated under the law of any part of the United Kingdom.[231]

Although the normal force of a 'declaratory' Act is to make it retrospective in operation,[232] if the argument rehearsed above is correct, this subsection reduces to statutory form what would otherwise have been the law since the coming into force of the regulations under section 3 of the Continental Shelf Act 1964. If this is the case, foreigners working on installations in the British sector of the continental shelf are subject to the criminal law relating to that part of the shelf within which the rig is operating, and are covered by the civil law to the same extent as their British colleagues.

C. The Application of Labour and Social Legislation

The extent to which labour and social legislation can be interpreted to apply to offshore oil-rig workers depends partly on the restrictions imposed by section 3 of the Continental Shelf Act, and partly on the nature of the legislation itself.

In considering the problem, an essential distinction must be drawn between those laws which can only have an effect on the installation itself, i.e. legislation relating to events at the place of work, and those which operate on the contract of employment and in other ways before the workers reach the installation. Thus safety legislation can only operate in respect of events on the rig, in which case its application is limited to 'acts and omissions' etc. This is not true, for example, of the Contracts of Employment Act 1972 which is related to the contract the worker has if he falls within its sections, no matter whether he is at work or not. Such legislation has nothing to do with 'acts and omissions' etc., and different considerations must therefore apply.

In line with the approach taken in this work, however, such legislation will be discussed at the point at which it is considered most appropriate. The Contracts of Employment Act 1972, for example, is best discussed with the details of the various contracts the Redundancy Payments Act 1965, with the other provisions relating to termination of employment, and so forth. As will be seen, specific provision has now been made for the extension of certain pieces of labour legislation to offshore workers,[233] enabling them to be discussed independently of the problems covered in this chapter. But the scope of the extension is not tantamount to blanket coverage of all offshore operations by all law in force onshore, and in respect of those laws which are not extended, reference must constantly be made to the question to which there is as yet no judicial answer: exactly what is an act or omission taking place in connection with exploration etc.? Consideration must then be given to the problem of whether the Act in question relates to acts or omissions, and if so, whether it does so in such a way as to include those occurring on installations. The completely absurd result which can be produced when legislation is looked at in this way is best discussed by an example: the Industrial Relations Act 1971 and the Trade Union and Labour Relations Act 1974. The solution proposed by the Employment Protection Act 1975 can then be fairly examined.

It is difficult to imagine that, other than with the most liberal judicial construction, the Industrial Relations Act 1971 was in any way concerned with 'acts and omissions' which could be said to occur 'in connection with' the exploration of the sea bed. The same is now true of the Trade Union and Labour Relations Act 1974. If this is the case, section 3 of the Continental Shelf Act cannot be said to have extended the application of that Act to workers on the continental shelf beyond the limits of territorial waters.

This is a problem which occupied the attention of certain of the offshore oil companies. Indeed, Professor Grunfeld, then Legal Advisor to the Commission on Industrial Relations, advised Shell that the 1971 Act was not applicable.[234] He took the view that the tortuous language of section 3 of the Continental Shelf Act would not have been necessary had the intention been to legislate for the continental shelf as if it were part of the United Kingdom, and that consequently it did not operate to extend general statutory or civil law provisions. But that approach, as already indicated, does not take into account the distinction between sections that apply to the employee *qua* employee, and those that apply to situations specifically outside the ambit of section 3. With this in mind one can re-examine the effect of these Acts.

The Industrial Relations Act purported to extend to 'every employee' and 'every worker'.[235] Needless to say, neither of these terms was defined in section 167 with reference to working on the continental shelf. Nor did the Act make any specific mention of any possible extraterritorial effect. This omission is repeated in the 1974 Act, but of course the Trade Union and Labour Relations Act contains far fewer provisions relevant in this connection.

In relation to the right not to be unfairly dismissed, however, section 27 (2) of the Industrial Relations Act, now paragraph 9 (2) of Schedule 1 to the Trade Union and Labour Relations Act, provided that the relevant provisions were not to apply 'to any employment where under his contract of employment the employee ordinarily works outside Great Britain'. A similar exclusion was to be found in section 57 (9) (c) of the 1971 Act, which denied the same category of employees the right to receive a financial statement from the company under that section.

Although the word 'ordinarily' presents problems of its own,[236] what is significant about these exclusions is that they should have been inserted at all. Two alternative interpretations are possible. Either the sections are clarificatory, and the Acts would not apply in any case to workers outside the United Kingdom. Alternatively the Acts would apply to workers whose contracts of employment were made in England, governed by English law, but to be performed abroad. The latter interpretation is preferred here. The relevant subsections act to exclude workers who would otherwise be covered. If this is the case, if the employee does not 'ordinarily' work outside Great Britain he could avail himself of the provisions of the Trade Union and Labour Relations Act relating to unfair dismissal if they were in all other respects applicable. On the other hand, if excluded it would make no difference if the worker were dismissed within Great Britain.

But the exclusively territorial effect of the Trade Union and Labour Relations Act has, in this respect, one further important result. Of those offshore workers covered by the Act, only those who were actually *dismissed* in this country were entitled to the protection of the statute, unless the dismissal could be seen as an 'act in connection with exploration etc.' Thus, instant summary dismissal taking place on any installation was outside the statute. Dismissal with notice which expired on land was covered in accordance with paragraph 5 (5) of the Schedule to the Act. Altogether a classic escape from the strictures of the law was provided for companies wishing to avoid the Act.[237]

The central theme of the above argument is that, unless the presumption against extraterritorial application is successfully rebutted, the statute only applies to events in this country. This adds a further perspective to the distinction between sections relating to the worker in person and those relating to the activities of workers. Clearly a worker engaged in this country, even for the purpose of working outside Great Britain, is entitled to benefit from legislation whilst within the jurisdiction. Thus, for example, the right given to every worker by section 5 (1) not to be excluded or expelled from membership of a union by way of arbitrary or unreasonable discrimination is one which is exercisable by the worker whilst in this country. Indeed, as long as the act of expulsion or exclusion takes place in the United Kingdom the section applies. But the oil-rig worker would not be able to complain were he discriminated against if the discrimination took place on the installation (it being outside the limits of territorial waters).

On the other hand, as contracts made in this country can be sued on in respect of breaches taking place abroad, rights embodied in a legally binding collective agreement made in accordance with section 18, or taking effect through the individual contract of employment, are enforceable out on the continental shelf.

The more difficult question relates to industrial action. Section 3 (2) of the Continental Shelf Act 1964 purports to apply the civil law relating to torts to events on the continental shelf. Does this apply the limited protection against actions in tort conferred by section 13 of the Trade Union and Labour Relations Act 1974 to action taken on an offshore installation in a designated area? The answer would appear to depend upon whether the act of inducing a breach of a contract of some kind out on an installation was an act 'in connection with the exploration of the sea bed etc.', and would thus appear to depend upon its purpose. Certainly the definition of a trade dispute in section 29 (3), which before 1976 did not bring workers in other unions within the golden formula if they called a sympathy strike over action on an offshore oil-rig (unless they were 'likely to be affected. . .by the outcome of the dispute' in respect of one or more of the matters set out in section 29 (1)) did appear to cover action by oil-rig workers themselves, at least on shore. But what about action offshore? The most logical answer appears to be that if the cause of the dispute was one which fell within section 3 (2) of the Continental Shelf Act 1964, then the question did not arise.

This short discussion should be enough to show that not enough consideration was given in the past to the application of the law to the continental shelf. The approach taken in the Continental Shelf Act, which was basically designed to extend the criminal law and the law of torts, is absurd in its application to wider legislation, and a far cry from the straightforward manner in which several other countries have simply extended the law, without distinction as to its classification, to their areas of the continental shelf. Nevertheless, as a result of representations made to the Department of Employment along these lines, Michael Foot promised in the committee debate on the Trade Union and Labour Relations Bill that future legislation would pay some attention to the legal problems of those working on offshore rigs,[238] and an effort has since been made in the Employment Protection Act 1975 to extend to them the coverage of some of the major pieces of labour legislation.

Section 127 (1) of the Employment Protection Act now reads:

Her Majesty may by Order in Council provide that the provisions of –

(a) the Wages Councils Act 1959;
(b) the Industrial Training Act 1964;
(c) the Redundancy Payments Act 1965;
(d) the Contracts of Employment Act 1972;
(e) the 1974 Act (i.e. the Trade Union and Labour Relations Act 1974 – see s.126 (1);
(f) this Act; and
(g) any legislation (that is to say any enactment of the Parliament of Northern Ireland and any provision made by or under a measure of the Northern Ireland Assembly) for the time being in force in Northern Ireland which makes provision for purposes corresponding to any of the purposes of any of the Acts mentioned in paragraphs (a) to (f) above, shall, to such

> extent and for such purposes as may be specified in the Order, apply (with or without modification) to or in relation to any person in employment to which this section applies.

By subsection (2) the section applies to work in the territorial waters of the UK or work in a designated area connected with the sea bed or subsoil or the exploitation of their natural resources, and the rest of the section provides for various ancillary matters designed to provide the customary flexibility. Thus an Order under section 127 (1) may make different provision for different cases,[239] or may apply to 'any person' employed for the relevant purposes in 'any part of the areas' set out in subsection (2).[240] The Order may provide that all or any provisions of any of the Acts mentioned in subsection (1) shall apply to foreigners or to foreign corporations,[241] and can determine whether or not any consent is necessary before any prosecution is brought under any of the Acts.[242] Obviously this section goes a long way to meeting some of the criticisms discussed above, and it should be noted that recognition of the limitations of the inapplicability of the language used in section 3 of the Continental Shelf Act is found in subsection (3) (c) which enables provision to be made for conferring jurisdiction on specified courts or tribunals 'in respect of offences, causes of action or of matters arising in connection with employment to which this section applies'.

This section of the Employment Protection Act, which echoes similar sections found in the Health and Safety at Work etc. Act 1974[243] and the Sex Discrimination Act 1975[244] both of which are discussed at more relevant places in this work), substantially extends labour law to operations on the continental shelf, particularly as an Order has now been made to give it effect.[245]

Under the Order (which does not apply to the area of the continental shelf in which the law of Northern Ireland is applied, nor, indeed, to employment 'wholly or mainly' for the purpose of any activities connected with the 'Frigg' field reservoir or the 'Ekofisk' field) all six of the Acts set out in section 127(1) of the Act are extended in their application to employment on all structures (including ships) in any state of construction or operation for use offshore for purposes connected with the exploration of the sea bed or its subsoil or the exploitation of their natural resources. The industrial tribunals of England, Wales and Scotland are given jurisdiction over matters arising in connection with employment from acts or omissions taking place in the relevant designated areas.

In line with the power set out in section 127 (3) (g) of the Act a number of modifications and exclusions of certain sections of the relevant Acts are made by the Order. For the most part these are confined to the extension of the meaning of the term 'Great Britain', e.g. in section 17 (1) and (2) of the Redundancy Payments Act 1965, to include territorial waters and waters in designated areas. Apart from this, the application of the Wages Councils Act, the Industrial Training Act and the Contracts of Employment Act is straightforward. The Redundancy Payments Act poses few problems, and, indeed, mariners not covered by the exclusion Order made under that Act in relation to merchant seamen are also included as section 20 is not applied. The provisions of the Trade Union and Labour Relations Act relating to unfair dismissal are modified in a similar respect so that seamen working on installations registered as ships are also covered. The extension of the Employment Protection Act itself poses rather more problems, but in substance these amount to no more than the fact that a further Order will have to be made at some point to extend the provisions of the Act not in force on 21 June 1976, the date the Order came into force. These include sections 22-28 (guarantee payments), 57-60 (time off work for trade union and other activities), and 98 (which with Schedule II is to replace section 8 of the Terms and Conditions of Employment Act 1959). That this has to be done at all reveals the inherent weakness of the approach adopted by the legislation. Although the extension and the extension Order represent a step forward for labour law and the employment of workers on the continental shelf, it is a somewhat uncertain one, or could be. The fact that the extensions are not automatic is a cause for regret as it makes the rights covered by the section subject to Ministerial discretion and hence to possible outside pressure in an industry in which there is already a large degree of potential government-industry conflict. It should be noted too that some important pieces of labour legislation, such as the Race Relations Act 1968, remain outside the section, and require still further parliamentary action.

3 RECRUITMENT TRAINING AND COMMAND

'What one can learn on the job and who will teach it depend on contingencies unrelated to education or training. The learning situation. . .is subject to. . .constraints emanating from the external world.' (Howard Becker, in *Learning to Work,* ed. Blanche Greer, Sage, Beverley Hills/London, 1972, p.103.)

Unlike the other offshore industries, fishing and shipping, the offshore oil and gas industry is not only unorganised by trade unions, but is so new, at least to the United Kingdom, that few rules or regulations have been produced to shape entry into the industry, work content or training. Thus until recently the industry was able to operate free of many of the restraints imposed on employers ashore. As time has passed and offshore oil has been placed more and more in the public eye, government concern has extended to the labour force in measures designed to increase safety at work, and to reduce discrimination in recruitment and employment. But before turning to the legal provisions which touch upon these areas it is necessary to describe the various different types of installation upon which an offshore oil-rig worker might be called to work, and the various tasks for which people are recruited, for, as V.L. Allen has said, 'one cannot abstract a worker from his environment and expect to understand him properly'.[1] This can be taken to mean not only the broader aspects of industrial behaviour and government response discussed in Chapter 1, but also the place in which he is expected to function.

I. RIG AND CREW STRUCTURE

Generalising about the structures of industrial establishments is inevitably difficult. In the case of the offshore oil industry, however, it is somewhat easier because, although the requirements of each company and the size of crew vary from rig to rig, it is possible to divide the crews up into broad functional categories, and the different rigs into broad types, any combination of which gives a good indication of the personnel involved. Installations can be classified in several ways for these purposes: into floating rigs and fixed platforms; into those which are stationary and those which are moving to another site; and into those which are concerned with exploration drilling and those involved in production.

The crews can be divided into compartments which match the various divisions of rig activities. The main categories are, broadly speaking, as follows: the drilling crew, which is concerned with the operation of the rig, as distinct from the platform which is managed by the barge engineer and his staff or by marine engineers or simply the ship's crew in the case of drill ships; the production staff; and the remaining members of the crew who divide into catering staff, the radio operator, the rig clerk who is an administrator, other ancillary workers, and the specialist engineers and supervisors who are attached to either the drilling or production sides.

Before going on to consider the composition of these divisions in detail three further factors must be mentioned. Firstly, the members of a rig are also divided horizontally by status as well as vertically by occupation. This will be apparent when the chain of command in each section is analysed. But there exists a broad division in almost every company[2] between the 'professionals' and the 'non-professionals'; between those who have had specialist as well as industrial training, university or professional qualifications, and those who have had 'on the job' training alone, who are employed on unskilled and semi-skilled jobs, and whose basic qualification is experience.

The second factor, which is connected to the first, is the shift system. Most of the North Sea operating companies operate a four crew shift system based on a twelve hours on/twelve hours off day, for either fourteen days on/fourteen days off, or seven days on/seven days off.[3] This means that for every position on the rig there are four workers, though only two of them will be on the rig at a time. But it also means that the accommodation has to be for twice the number of people actually at work, which on a small installation can be of considerable importance.[4] In fact, the shift system does not extend to the professionals in quite the same way. These workers may well alternate on the rig on a weekly basis, each one working for twelve hours, then being on call for the other twelve. Alternatively, the company may operate a four man shift system of twelve hours and seven days. Generally speaking, when on the rig they will be expected to work a full twenty-four hours in the event of difficulties.

Lastly, the Mineral Workings (Offshore Installations) Act requires a member of the crew to be the designated 'Manager' of the installation.[5] This is discussed in detail below.

II. JOB CLASSIFICATION AND DESCRIPTION

Following the distinctions made between the various types of rigs, the crew themselves can be discussed under their various compartments of activity.

A. Drilling Crew

Most of the rigs mentioned in this study are engaged in drilling, i.e. boring and casing the hole down to the oil or gas, either as exploration rigs only doing drilling work, or as production platforms doing field development drilling, i.e. drilling further holes down to a known field for development and production purposes. The actual drilling operations are performed by the workers on the rig (strictly defined as the tower and drilling apparatus, as distinct from the more general use of the word which stems from the land situation where the rig is not supported by a platform). The man in charge is known as the 'toolpusher', and he will often have an assistant, and will normally have an alternate. He is well recognised in the industry as being in overall control of operations and the installation when engaged on drilling.[6] He is a driller by training, which is specialised in the sense that companies run courses for drillers to bring them up to date, and the person appointed as toolpusher is unlikely to have less than five or seven years' experience as a member of a crew. Most of the men in the North Sea who occupy this position have been brought over from the United States or other parts of the world with a longer history of offshore operations.

Below him will be the driller, and an assistant driller. Their principal task is manipulating the controls on the drilling machinery to keep the drill going down at a constant speed. The driller also has to adjust the weight on the drill (which to a certain extent determines the downward speed) caused by the actual weight of the pipe behind the bit. If it is too high the hole will run crooked and the pipes will bend under their own weight. At the same time too much counterweight from the lowering machine will mean the bit is not descending. He also has to be able to 'feel' the progress and be on guard for gas and oil, watching constantly to prevent a 'blow-out' – a powerful upsurge of gas or oil through the hole caused by release of pressure at the first strike. If the blow-out preventer (a complicated system of valves designed to close around the drill string or close off the hole if the upward surge cannot be contained by the mud) fails, the erupting oil or gas can completely destroy the equipment, or the rig, and is a grave fire hazard at all times.

It is possible to be promoted from the rig floor to the position of assistant driller, though this is rare, and men are usually recruited directly with HNDs or the equivalent[7] and then have the usual lengthy[8] on-the-job training and special courses. In many cases these men will be brought from other parts of the world because of lack of expertise in Britain.

Below the specialists are the semi-skilled employees. These are derrickmen who work in the rig tower; floormen, who work for the driller taking the bits of the drill and handling the pipes put down to case the hole; and the crane operator, whose job is central in the sense that the crane moves the pieces about the rig in all weathers, and is therefore extremely dangerous. Drilling crew are known in the industry as 'roughnecks'. The manual workers who do the dirty handling jobs and general labouring are known as 'roustabouts', and work as part of the drilling crew.

Entry into the industry is usually as a very green 'roustabout'. These men are unskilled, and for the most part 'just pick it up' as far as learning their job is concerned. They are usually recruited locally, though some come some distance in response to advertisements that are placed in the press. Indeed, many companies prefer to take on men who have been recommended to them by other workers in the company – so entry is often via a friend. It is in fact one of the most remarkable features of the industry that, at least until 1974, the companies had little trouble in recruiting, even though the work is dirty, dangerous and hard, involves living in an enclosed space for a fortnight at a time and is subject to rigid discipline. Both British Petroleum and Shell had lists of people who have written to the companies to apply for jobs without any prompting at all. Some companies, such as ODECO, get 'walk-in' applicants, and a certain number are found through the Employment Exchanges. There is some evidence, however, that certain companies have been having more and more difficulty recruiting since 1974. The reasons are unclear, but one might speculate with some cause that increased public awareness of the dangers involved, and the discovery that the pay is not infrequently much lower than expected, are contributing factors.

As far as employee profiles are concerned, it seems that because of the constraints of the job, ex-seamen and ex-servicemen are regarded as good recruits. Shell stated that they only employed 'the better kind of individual'. There is evidence too that some companies have given preference in the past to non-unionists. There is one company, Notecon, who provide 'labour only'. This company hires roustabouts to operating companies in gangs. There is no evidence, however, that 'the lump' has been widespread.

B. Platform Personnel

On a floating installation there will be personnel concerned with the installation itself. They will be under the direction of the head or senior barge engineer, who is in charge of the stability, ballasting and trim of the installation. When the rig is being moved from one site to another he will take over control from the senior toolpusher. He may have a deputy on a large rig like the Shell semi-submersible 'Staflo' or the 'SEDCO 703'. He will have a crew of one or more senior or assistant rig technicians, a motorman, who is in charge of the engines, and mechanics, welders, and electricians. The mechanics and engineers are all recruited as specialists with the usual training, though they will all undergo further specialised offshore training under the company. Reading and Bates[9] send their electricians to be trained in the USA in Houston with their parent company. These workers are included under the heading of 'platform personnel' because they are attached to that division of the crew when one exists. There may well be one or more engineers, welders and electricians on a fixed or jack-up rig, in which case they will come under the crane operator.

C. Production Staff

On a production rig the man in charge of actual production will be the production supervisor who is recruited with qualifications (usually HND's, but may have been promoted from one of the natural gas technicians, who are other specialist production staff. This sort of crew will also contain electricians, mechanics, fitters, lab-assistants and the usual catering staff.

D. Specialist Personnel

Apart from the toolpusher who remains on the drilling rig, there are always a number of specialised personnel who will visit the rig from time to time and in many cases will be periodically attached to a rig for a substantial period of time. The principal types are petroleum engineers, mud engineers (who are nearly always permanently on a rig), geologists and geophysicists. The mud engineer is a specialist who is concerned with the technicalities and problems of the mud which is forced down the hole to lubricate the drill bit at the point of drilling and contain the pressure when oil or gas is encountered. The petroleum engineer's task is a difficult one. He has to bring up the pieces if the drill breaks (though this will sometimes be done by 'fishing' tool specialists), cope with the gas zone problems, take logging runs, and do tests. He will sort out the chippings that come up and also oversee the drilling programme in conjunction with the drilling supervisor (the toolpusher) for casing the hole and ensuring that the drilling programme is met. In some cases he might well be in overall command of the installation. All (like most toolpushers) will have engineering degrees, and have done specialist courses and supplementary courses to bring them up to date.

These professional staff will all be recruited centrally, and will generally operate anywhere in the world for an international company, moving within a centrally determined career plan. A Shell petroleum engineer, for example, will join the company with a degree and then do a six-month specialist course in the Hague. He will then join an operating company (such as Shell (U.K.) Exploration and Production Ltd) for a year's work on a rig. He will then do a spell as an assistant to a specialist for about nine months, then return to the Hague for a supplementary course. After this he is ready to take a position as a junior specialist. This pattern is followed by most field engineers, although British Petroleum send their recruits to the United States for familiarisation trips. Many will find themselves in most of the offshore oil fields in the course of their careers, and they will move when a job ceases, or when they are due for promotion. In most companies these men are foreign nationals, usually American, though some will be Dutch in the Shell organisation.

E. Ancillary Staff

Apart from the specialist staff all rigs will have a certain number of ancillary personnel. These will include the radio officer, who will nearly always be on loan from Marconi to the operating company; the rig clerk; divers, and the catering staff. Divers are a specialist breed who well merit a study of their own.

Some are employees of the operating company (for example, British Petroleum employs some divers of its own) but most are the employees of independent diving companies. Shell divers are on standby from a company half owned by Italian Shell. Most are employed by specialist diving concerns providing subsea services to drilling companies, pipe-laying companies and so forth. The largest single source of recruits has been the Royal Navy, but with rising demand for divers' services a good number of less well trained men have joined independent companies after basically only amateur experience. This situation is causing a good deal of concern, and training schools are opening and standards are being promulgated to rectify these problems.

Catering staff, too, are employed by outside firms who tender for contracts to supply installations with full catering and housekeeping for a fixed price for a specific number of people. They provide the food and other incidental equipment. The employees enter the trade in the usual ways for the catering industry. Thus cooks will have already received the same training as cooks in good hotels or catering concerns, generally involving training courses at technical or catering colleges; and additional on-the-job directed experience. There may also be instrument technicians, recruited from the electronics industry.

F. Seamen and Pipe-layers

If the rig is mounted on a drilling ship, somewhat different considerations apply. Those of the crew concerned with navigation, and probably the catering crew and some of the maintenance team as well, will be seamen employed in the traditional manner as members of the shipping industry and recruited in the normal ways in that industry. On the other hand, those who operate the drilling tower and the various specialists will fit the categories described above, and the same considerations will apply.

It is difficult to generalise about the crews of pipe-laying barges, which are of a number of different types. The lay barge is the kind most used in the North Sea. These barges haul themselves forward by a series of anchors and motors, and the pipe, which is welded together on the deck from sections made ashore, is laid behind the barge in a pre-cut trench as it moves forward. Obviously welders, crane operators and divers are needed on these operations, as well as those concerned with navigation and the movement of the equipment and ancillary crew such as storekeepers, mechanics, maintenance men and catering staff. Most will be recruited in the same way as they would be if they were working on an actual rig, and can be treated much in the same way except that it must be remembered that they are employed on 'pipe-line works' and not on 'offshore installations', and are therefore subject to different legal considerations if this distinction is of relevance.

G. Overall Personnel Requirements

In their Report the International Management and Engineering Group examined what they thought to be the personnel requirements for an exploration rig in operation.[10] The Report concludes that some sixty-five people are needed on each rig, including ten professional staff of one kind or another such as three drilling engineers and three marine engineers, some six technicians, twelve members of the catering crew and over thirty non-professionals. It should be noted that the Report was examining the theoretical needs of such a rig rather than documenting normal practice.

In fact, most rigs on location in the North Sea will have a crew of around seventy, although some of the bigger semi-submersibles such as the 'SEDCO 703' have more, and, of course, there are a number of smaller rigs carrying fewer personnel. For example, on some rigs Reading & Bates have about eight catering crew, twenty-eight shift crew and the total number of persons on the rig at any one time will probably only be about fifty when the specialists, ancillary staff, the radio operator and others are included. The British Petroleum rig 'Sea Quest' on the other hand, has a crew of about seventy, with about thirty men working on each shift. The Shell rig 'Staflo' has a crew of just over fifty, of which about forty are shift workers. But it should be remembered that each crew requires duplication to take account of the shift system. If a rig can only operate with thirty non-professional workers on board, the overall number of these workers for that rig will be sixty. Any variation between the needs of the various rigs is therefore multiplied, and it should be recalled that the crew demands of a rig doing development drilling as well as production, and a simple production platform, will be enormously different. As a result it is difficult to generalise with any great accuracy about precise manning requirements, or to document exactly who is needed on each rig.

Some overall figures of the numbers of people working offshore are available, however. Excluding seamen working on attendant vessels and people working on pipe-laying barges, reports to the Department of Energy show that the total number of persons employed in 1974 was 4,030. During the same year there were twenty-three production platforms in operation, and drilling activity in terms of rig years amounted to 24.5 for mobile rigs, and 2.8 for fixed platforms.[11] On a straight average basis this would confirm much of what has been said here, but it should be emphasised that exact figures for the numbers of persons employed are, on the Department's own admission,[12] nearly impossible to obtain. There is considerable difficulty in taking account of short visits, and labour turnover and the fact that the particular number of rigs operating in the UK sector of the continental shelf changes over the course of the year with new ones moving in, and others moving over the median lines or ceasing operations, complicates the picture considerably. With an increased number of production rigs being brought into operation this situation will stabilise somewhat. Nevertheless accurate assessment is likely to remain difficult, and the overall numbers employed will be subject to fluctuations depending on the success of the drilling programmes, future discoveries, the time of year, government policy, and the overall economic climate within which the operations are taking place.

III. TRAINING AND LABOUR TURNOVER

As training inevitably requires a certain amount of investment in the labour force it is not surprisingly related in certain circumstances to labour turnover. In sharp contrast to those industries where training is a formal requirement, training in a more loosely structured industry like the offshore oil industry can in certain cases have an enormous impact on personnel behaviour and motivation. Thus, where there is labour turnover training can act as a stabilising factor, particularly if the resulting skills are specialist and the training has status attached by the company in the way of reward.

In the offshore oil industry there is evidence of a reasonably high turnover, particularly amongst roustabouts,[13] which in the case of at least one company has periodically reached 40 per cent of each crew per fortnight on some rigs. Although no accurate figures are readily available, every company consulted in the course of this study pointed to this as their only labour problem. It appears that the turnover is particularly high in the summer months when the casual nature of the employment ostensibly enables workers to earn enough money to take several weeks' holiday and then find employment with another company on the basis of previous experience. The reasons for this phenomenon are not clear, although some hypotheses may be presented, drawing on factors which have been thought important in the shipping industry.[14]

Firstly, though the preponderance of the crews of some companies will be British,[15] working offshore means a break with the family. Some 50 per cent of the crews of rigs run by Reading & Bates are estimated to be married. Separation from home for two weeks out of four or for every other week is something normally undertaken only by young workers.[16] According to Reading & Bates some 80 per cent of their crews are under thirty years of age, and 50 per cent are under twenty-five. The minimum age for employment under the law is now eighteen, and the company does have workers of that age. Increasing age tends to bring a desire to be closer to home, especially if there are children in the family.

Secondly, whereas it is possible to 'see the world' in the shipping industry,[17] this cannot be an important factor in the desire to work on an offshore rig. For this reason there is a certain interchange of personnel between companies of workers who want to stay in the industry but would like a change of 'scenery'. The fact that turnover is lowest in winter is in line with the diminished employment opportunities with other companies.

There are, of course, other contributing factors. It is not possible to form a realistic assessment of the job without visiting a rig, and no companies provide for potential recruits to 'look over the rig' before being hired. Labour turnover may therefore simply reflect an awareness of the dislike for the work or the environment realised only during the first shift. The work might turn out to be more dangerous than the recruit imagined, harder, subject to more rigid discipline, and frequently far more frightening. Again there are parallels in the labour turnover pattern of those going to sea in ships, a large number of whom leave after the first voyage.[18]

The companies have therefore turned their minds to methods of holding the workforce. Some companies run incentive schemes. But the week or fortnight ashore provides a good opportunity to look for alternative work, and in many cases is necessary because some of the companies, apart from Shell and British Petroleum, do not pay for that week. It is a frequent complaint that up to 40 per cent of those who leave do so by failing to turn up to catch the helicopter at the start of the fortnight.

As training a man in specific skills plays an important part in reducing incentive to look for another job, it is not surprising that it is particularly amongst roustabouts that labour turnover is considered a problem. For the most part they receive little training, except that gained on the job like all other workers. But there have been a number of recent changes in this situation, and a number of organisations have been created to cater for the training of oil-rig workers.

In many respects the interest was slow to develop. But in July 1971 the Petroleum Industry Training Board, set up in 1967 under the Industrial Training Act 1964,[19] formed the Petroleum Training Association (North Sea), known as PETANS, based on Great Yarmouth.[20] A separate Scottish Offshore Training Association, known as SCOTA, was also formed under the auspices of the PITB in the autumn of 1973,[21] and the North Sea chapter of the International Association of Drilling Contractors and the United Kingdom Offshore Operators' Association sponsored the Offshore Petroleum Training Association in conjunction with the PITB almost at the same time.[22] All these bodies have added to those facilities already provided by the industry itself. In fact one of the principal steps towards the provision of training facilities was taken by one of the leading drilling companies, South-Eastern Drilling Co. Ltd (SEDCO) which opened a training centre for drilling personnel in Aberdeen on 6 November 1972,[23] looking to provide basic training for unskilled workers entering the industry as roustabouts.

All these facilities have been created to meet a growing demand. The SEDCO centre expected to give instruction to about 500 employees in the first two years of its existence, and some future projections of the numbers needed to supply the offshore industry in the next decade have made these figures look fairly small. The IMEG Report devoted some considerable space to training[24] and concluded that the Government should purchase a jack-up rig (at a cost of about £4 million) for training purposes.[25] Heavy investment has also been recommended by two more recent reports. An inter-departmental working party set up under the auspices of the Department of Employment in May 1973 produced extensive projections of the personnel needed by the industry (reproduced as Table 1 below) and was concerned that further developments in training should match this kind of demand.[26] A special report prepared for the Manpower Services Commission in 1974[27] also recommended the urgent establishment of a deep-diving training centre in western Scotland at a cost of about £2 million. Other specialists have also been in short supply, and several universities have instituted courses to train geologists and engineers for offshore work. In the light of the slow-down in the pace of development and production, however, there is some danger that the predictions of future needs might have been optimistic, although expansion of the industry throughout the world would absorb the professionals, and labour turnover might well do the same for the unskilled and semi-skilled, many of whom might also be easily employed in other parts of the industry, e.g. in the development of onshore facilities and terminals.

Table 1. Projected Personnel Expansion: Training Needs 1973-80

	1973	1976	1980
Non-professional			
Drillers	160	250	320
Crewmen	1020	1690	2550
Mechanics/Fitters	550	950	1600
Electricians	280	480	800
Communications/Instruments	150	340	630
Plant operation	160	350	650
Lab Technicians	40	80	110
Divers	200	340	480
Chef	220	380	610
Stewards	370	640	1070
Radio operators	120	210	340
	3270	5710	9160
Professionals			
Drilling Technologists (Supts., Toolpushers, Drill engs.)	230	350	450
Petroleum Engineers	110	230	370
Geologists	80	120	150
Geophysicists	20	30	40
Mechanical Engineers	120	210	340
Electrical/Electronic Engineers	160	310	500
Instrument Engineers	20	50	80
	100-120	130-170	180-260
	860	1470	2190
TOTAL	4130	7180	11350

Source: *Education and Training for Offshore Development,* Report of an Inter-Departmental Working Party, Department of Employment, 1973, Annex 4.

In fact an increasing number of courses are being made available. PETANS has run courses in offshore crane operating providing basic training for these key workers since 1972.[28] The courses have been designed to give instruction in all aspects of crane use. It is in three stages: a basic onshore course; offshore evaluation and loading under supervision; and a period of on-the-job planned experience under the supervision of company management. PETANS also run courses on other subjects such as 'blow-out' prevention, fire-fighting, safety and first-aid.[29] Both PETANS and SCOTA have put on courses relating to helicopter escape training, survival (based on the use of safety capsules and life rafts), and the special responsibilities of installation managers. SCOTA has further plans for fire-fighting and rescue training facilities, and other branches of the PITB have arranged courses on seismic data acquisition, processing and interpretation.[30]

In particular, basic induction courses for recruits have been developed by the Board, and are run not only at the SEDCO centre, but by the OPTA in Aberdeen. In fact a new Drilling Technology Training Centre has been established at Livingston near Edinburgh by the PITB funded through the Manpower Services Association and the OPTA sponsors, although it will probably move nearer to Aberdeen when it acquires a dry training rig. After pilot courses had been run for deep diving under the auspices of the Ministry of Defence (Navy) Marine Service Diving School, the Government gave permission at the end of 1974 for the establishment of the Deep Diving Training Centre, recommended by the Manpower Services Commission Report. The centre has since been established by the Training Services Association and Shenley Trust Ltd on Loch Linnhe near Fort William, and the first intake of ten recruits took place in September 1975.

All these various establishments will undoubtedly fill an enormous need bearing in mind the special dangers of the peculiar working environment. It must be borne in mind too that many workers will have received their specialist training ashore under well-established conditions, e.g. fitters and welders who serve an ordinary apprenticeship, and radio officers who qualify as they would to work on ships. Others, including specialists and divers, will have received their training abroad, particularly in the United States where the offshore industry has been established for some time.

IV. DISCRIMINATION IN EMPLOYMENT

It is appropriate here to mention both the application of the Race Relations Act 1968 to employment on an offshore installation,[31] and the potential application of the Sex Discrimination Act 1975. The former is particularly relevant since workers of many different nationalities are employed in the industry.

A. The Race Relations Act 1968

Section 3 of the Race Relations Act makes it unlawful for an employer to discriminate[32] against a person seeking employment, either by omitting to employ him on available work for which he is qualified,[33] or by failing to give him similar terms and conditions of employment or opportunities for training and promotion as those he makes available to persons of similar qualifications employed in like circumstances on work of a similar description.[34] But its application to coloured workers, or indeed British workers, seeking employment on the continental shelf, is beset with problems.

Section 8 (7) of the Act reads:

> Those sections[35] shall apply to any employment which is, or an application for any employment which is to be –
> (a) wholly or mainly in a country outside Great Britain; notwithstanding that the person employed or seeking employment was engaged or applied for it in Great Britain.

As the continental shelf is not a '*country* outside Great Britain', it would appear that, construing the statute rigorously, the Act would apply to the engagement of workers in this country. The problem arises with ensuing paragraphs of the same subsection of the Act which exlude from the coverage of sections 2 and 3 employment on ships, British or otherwise, outside Great Britain. As some drilling rigs, such as the 'Gomar IV', are registered as ships, it would appear that the Act only applies to those rigs which are not so registered.

An even more absurd result is nevertheless contained in the legislation. Section 6, which makes unlawful any advertisement or notice which indicates or could reasonably be understood as indicating, an intention to discriminate, applies to companies and events in Britain, but not in relation to Commonwealth citizens required for employment outside Great Britain. Furthermore, acts of discrimination which take place on the installation itself, outside Great Britain though on the continental shelf, are not susceptible to enforcement through legal proceedings.[36] Thus if section 3(1)(c) applies at all, it does not apply to a dismissal out on the rig because it occurs abroad. In all cases the only remedy will be through the contract of employment if it is broken.

Once again the determination of the above argument has not been made by the courts. But, bearing in mind the arguments of Chapter 2 relating to the extraterritorial effect of statutes, it is submitted as a correct conclusion unless discriminatory acts on an installation could be called 'acts and omissions in connection with exploration etc.'. As the situation stands, however, the Act does apply to recruitment and engagement of offshore workers, to their terms and conditions of employment and to the provision of training and promotion. But it does not apply to acts on board the rig itself.

This conclusion is both regrettable and marked, for on some rigs, and on some supply vessels, separate living quarters are provided for workers imported from Spain and other non-north European countries. Although it is generally held that race is not a problem out on the rigs, and indeed the cosmopolitan nature of the industry would mitigate against it, some discrimination does occur. For this reason it is very likely that proposed changes in the Race Relations Act will include the power to extend the application of the legislation to work on the continental shelf. It should be noted that the Act was excluded from the extension section of the Employment Protection Act, highlighting yet again the difference between the method of applying British law to offshore operations, and the way other countries have simply extended the scope of law in general.

B. Discrimination on the Basis of Sex

It is virtually unknown for a woman to be employed in the offshore oil industry in much the same way as it has hitherto been extremely rare for women to be employed in the merchant marine. This situation might well have to change as a result of the passage of the Sex Discrimination Act 1975.

Under section 6 (1) of the Act it is now unlawful

> . . .for a person, in relation to employment at an establishment in Great Britain, to discriminate against a woman –
>
> (a) in the arrangements he makes for the purpose of determining who shall be offered that employment, or
> (b) in the terms on which he offers her that employment, or
> (c) by refusing or deliberately omitting to offer her that employment.

Furthermore, by virtue of section 6 (2), if a woman is employed it is unlawful

> . . .for a person, in the case of a woman employed by him at an establishment in Great Britain, to discriminate against her –
>
> (a) in the way in which he affords her access to opportunities for promotion, transfer or training, or to any other benefits, facilities or services, or by refusing or deliberately omitting to afford her access to them. . .

Obviously the limitation of the section to establishments in Great Britain would have the effect of making the Act inapplicable to those installations outside territorial waters. Indeed section 10 (1) provides that employment is to be regarded as being at an establishment in Great Britain unless the employee does his work 'wholly or mainly' outside Great Britain.

However, the Act makes it possible to get round this limitation. Firstly it provides that employment on board a ship registered at a port of registry in Great Britain is covered by the Act unless the employee performs his work 'wholly' outside Great Britain.[37] Unless the ship works out of a port in one of the other countries around the North Sea, this means that drill ships registered in Britain and working out of a British port would appear to be covered by the Act. But this might not mean that the workers on board would fall under the protection of the Act. They might be members of the drilling crew and only work outside territorial waters. They might be flown out to the ship to work, as indeed many are. In both cases section 10 (2) is of little help. But section 10 (5) provides that

> In relation to employment concerned with exploration of the sea bed or subsoil or the exploitation of their natural resources, Her Majesty may by Order in Council provide that subsections (1) and (2) shall each have effect as if the last reference to Great Britain included any area for the time being designated under section 1 (7) of the Continental Shelf Act 1964, except an area or part of an area in which the law of Northern Ireland applies.

Thus the provisions of the Act may be extended to offshore operations and indeed if an Order is not made the industry is once again faced with the absurd possibility of a law applying to those installations within territorial waters but not to those just beyond the limits.

However, at least one of the exceptions provided for in the Act might be applicable to installations whether or not they are situated within territorial waters. Section 7 (1) of the Act provides that section 6 (1) (a) and (c) and section 6 (2) (a) of the Act does not apply to any employment where being a man is a genuine occupational qualification, and this includes the situation where –

> . . .the nature or location of the establishment makes it impracticable for the holder of the job to live elsewhere than in premises provided by the employer, and –
>
> (i) The only such premises which are available for persons holding that kind of job are lived in, or normally lived in, by men and are not equipped with separate sleeping accommodation for women and sanitary facilities which could be used by women in privacy from men, and
>
> (ii) it is not reasonable to expect the employer either to equip those premises with such accommodation and facilities or to provide other premises for women. . .

In fact, of course, the relevance of this provision depends on the interpretation placed upon it by an industrial tribunal.[38] But power is taken in section 10 (6) for an Order in Council made under section 10 (5) to modify the provisions of the Act in their relation to employment on the continental shelf, and indeed it is unlikely that an Order would in fact be made without modification of this provision. Whether or not that means that such an Order will ever be made remains to be seen.[39]

Were women to be employed on the continental shelf it should be noted that the provisions of section 10 of the Sex Discrimination Act also apply to section 1 of the Equal Pay Act 1970, subsection (7) of which would restrict its operation to employment in establishments in Great Britain. As from 6 April 1978 this subsection will in fact be repealed,[40] eliminating all question of conflict between the two provisions.

C. Disabled Persons

A limitation on the applicability of the Disabled Persons (Employment) Act 1944 to employment in Great Britain[41] also takes employment on the continental shelf outside the scope of the Act if the installation is outside territorial waters, except where the establishment is a British ship and the employment is entered into in Great Britain.[42] There is no provision for the extension of this Act, and indeed it would be surprising, given the general approach of the legislators to employment on offshore oil-rigs, if any thought had even been given to the application of the Act to installations situated within territorial waters. It nevertheless seems unlikely that any amendments will be made.

V. COMMAND

A. The Industrial Command Structure

The hierarchical structure of command on board an offshore installation should be apparent from the description of the tasks to be performed. As in most industries it follows the status and difficulty of the job, and varies between the vertical divisions of the crew and the operation being performed by the installation at the time. Obviously the specialists will be allowed rather more independence than those other individuals who operate outside the groups which function as units on board, but overall responsibility for the allocation of work will reside with the person in charge of operations.

These points are perhaps best made by the following illustrations, which set out common command structures found on different types of offshore installation. Figure 1 (which sets out the structure on an exploration rig whilst engaged in drilling) and Figure 2 (which shows the structure of a similar rig while moving site) should be contrasted to show how responsibilities naturally change according to what the rig is doing. The section of the crew concerned with the platform, who are at work all the time on a semi-submersible ensuring that it remains in position, are naturally not present where the platform has been placed in a permanent position on the sea bed for production purposes (as in Figures 3 and 4) even where (as in Figure 3) development drilling is being carried out from the platform. In those cases, of course, the position of the structure will be frequently checked against scouring and sea bed movements, and for faults and wear and tear. It should be borne in mind when examining the Figures that not all the personnel indicated will be present all the time on some of the smaller rigs, and may also vary according to the state of operations.

In the case of chartered rigs, as distinct from those rigs owned and operated by the same company, the senior toolpusher will normally be the man in charge, and an employee of the operator, i.e. the company that has undertaken to be in charge of the operations on behalf of the licensee or the consortium of companies that holds the licence for the

Figure 1. Command Structure on an Exploration Rig whilst Engaged in Drilling

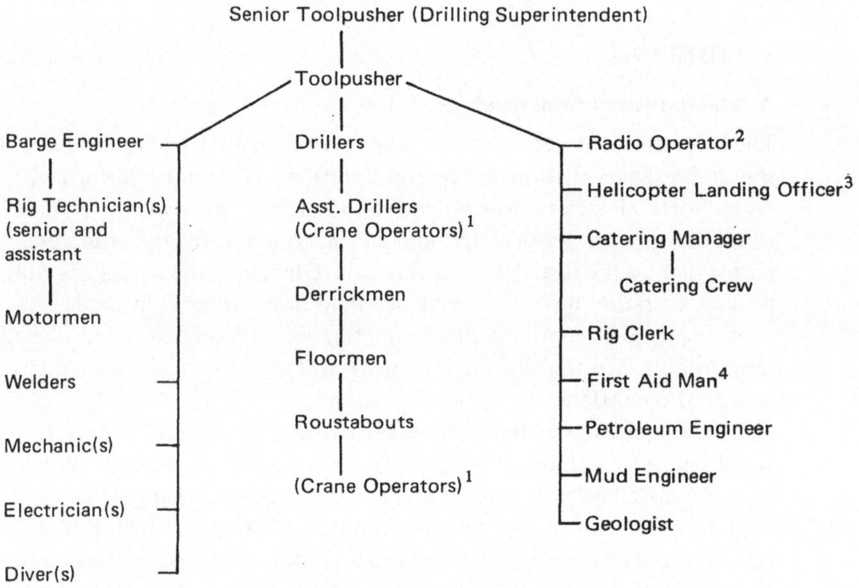

[1] The crane operator is accorded a different status in different companies, e.g. higher with Conoco, and lower with Shell.

[2] Radio operators are now obligatory on all manned offshore installations. See the Offshore Installations (Operational, Safety, Health and Welfare) Regulations 1976, S.I. 1976, No. 1019, Regulation 19.

[3] One person on each rig must now be designated the 'helicopter landing officer' in charge of helicopter operations. See S.I. 1976, No. 1019, Regulation 21.

[4] Every installation must carry at least one 'medically trained person' (two if more than 40 persons on board). See S.I. 1976, No. 1019, Regulation 31.

Figure 2. Command Structure of an Exploration Rig Moving Site

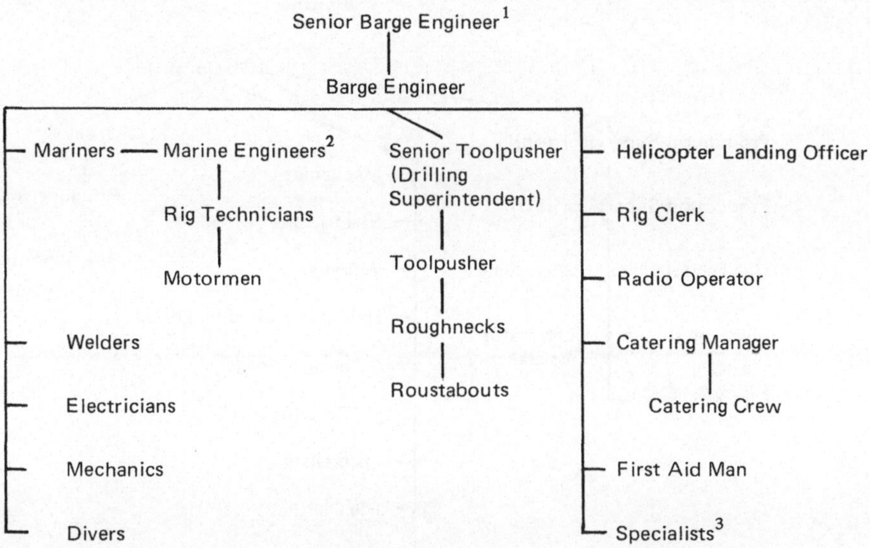

[1] The Senior Barge Engineer may come aboard to supervise the move.

[2] Marine Engineers will be brought on board to help pilot the installation when moving. In some cases, e.g. on the British Petroleum rig 'Sea Quest', they are on board all the time. Their name explains their job.

[3] If still on board.

Figure 3. Command Structure on a Combined Drilling/Production Rig

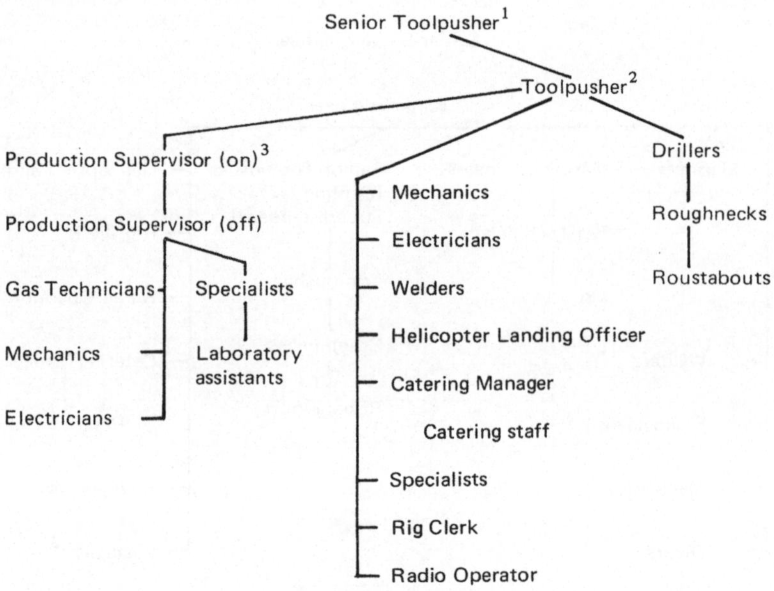

[1] The Senior Toolpusher is in overall command, and might be the operator's employee.

[2] The Toolpusher is in charge of the drilling crew, and ranks above the Production Supervisor.

[3] The Production Supervisor is in control of the production side, but although he ranks below the Toolpusher, will act as production adviser to the Senior Toolpusher.

Figure 4. Command Structure and Personnel of a Production Platform

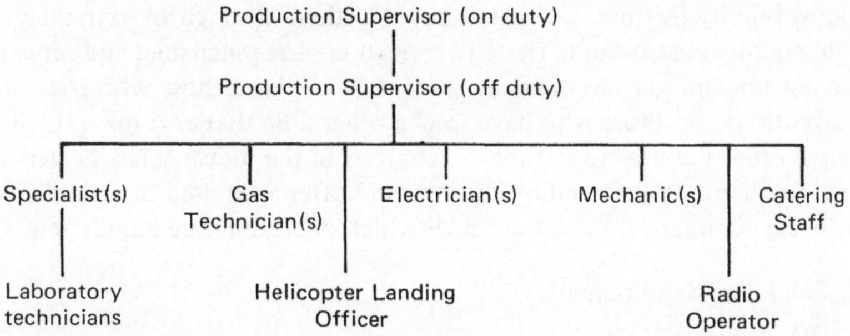

area in which drilling is to take place. Apart from specialists, who would also be employees of the operator, the crew would mostly be employed by the contractor, the company that hires out the rig for use, whose senior representative will be the toolpusher and overall second-in-command. The catering crew may also be employed directly by a sub-contracting specialist catering company providing services to the rig, and therefore under the immediate control of the catering manager.

In the case of those platforms which are straightforward production platforms not involved in development drilling (Figure 4), it should be recalled that if manned at all they are likely to have only a skeleton crew aboard, concerned mainly with measuring and controlling the flow of oil or gas to the shore and performing maintenance work. The total number of persons on board such an installation will rarely be above fifteen.

There is some difference of opinion as to the degree to which crew members function independently within the overall structures outlined here. In theory, at least, many of the tasks to be performed on the rigs are self-supervising, and relatively self-contained. Thus the mechanic looks after all the mechanical problems as they arise, and the electrician will tackle all the electrical work that needs to be done. The industry is of the opinion that this enables the crew to function very much as a team with common goals, and indeed one MP remarked rather idealistically in the House of Commons that 'The idea of a hierarchy of command, in a sense, does not exist. This is a team which does a job.'[43] On the other hand, those working as trade unionists in this area have been at pains to stress that this picture is far from correct, and indeed that many accidents would not have occurred had the workers concerned been allowed to use their common sense rather than being obliged to follow blindly instructions passed down to them through an extremely rigid command structure. These two views are irreconcilable, and reflect the not uncommon variations of perception between those who give instructions and those who have to obey them. Be that as it may, it will be seen that the position of the order-givers in the industry is a powerful one, frequently reinforced by the law, and often exercised in accordance with the commercial pressures under which operations are conducted.

B. The Legal Requirements

1. *The Manager*

One person on every installation has always been seen by those working on board as in charge of operations and of every other person on board. Nevertheless the Report of the Inquiry into the loss of the Drilling Rig 'Sea Gem' recommended that there should be someone designated as being in command at all times. There were two aspects to this recommendation.

The Institute of Petroleum Code of Safe Practice for Drilling, Production and Pipeline Operations in Marine Areas,[44] which is incorporated into the licences by instructions given under the Model Clauses authorised by the Petroleum (Production) Act 1934, and the Continental Shelf Act 1964,[45] states that '. . .the individual in full charge of all operations carried out. . .and to whom all personnel are responsible in an emergency, should be clearly defined'.[46] Although this was indeed the practice, the Tribunal were unhappy with the result for two reasons.

Firstly, though the Tribunal acknowledged that the toolpusher was regarded as being in full charge notwithstanding that drilling was not in progress[47] they found that in the case of the 'Sea Gem' there has been a frequent change of toolpusher on the rig. They asked whether this was 'conducive to a general feeling on the barge that a known and trusted person was in command'.[48] Secondly, they were unhappy that the toolpusher, selected for his ability as a drilling supervisor, might not have the necessary 'powers of leadership', although the industry has always seen him in that light and the company is hardly likely to have put him in that position without thinking him in all respects suitable. The Tribunal therefore concluded that

> . . .a named person and one whose powers of leadership are

> recognised should be in charge of all operations rather than the fortuitously allocated performer of a particular function related more to the technique of drilling than to the safety and well-being of the barge's whole company.

The second basis for the recommendation was an analogy drawn with the merchant navy. The Tribunal was of the opinion that at the time of the disaster: 'the pontoon had the character of a sinking ship, and the fact that it formed part of an oil drilling rig was irrelevant.'[49] As a result they finally recommended that:

> Although many of the perils from which those serving on drilling rigs must be protected are not in their essence maritime perils the fact that the 'Sea Gem' was lost in the character of a sinking ship suggests strongly that there ought to be a 'Master' or unquestioned authority on these rigs and that a second in command or 'Chief Officer' also be designated.[50]

This was made expressly 'because the condition in which (the Tribunal) has to view the 'Sea Gem' is that of a sinking ship and not as an oil-rig'.[51] Indeed they backed this up with the further recommendation that: '. . .there ought to be an accepted discipline and a chain of command analogous to what has been established over the years by the practice of the Merchant Service fortified by the provisions of the Merchant Shippings Acts.'[52]

These conclusions are open to several different criticisms. It does not necessarily follow that simply because work takes place at sea there is a definite need for a 'master', or that the provisions of the Merchant Shipping Acts (or something similar) should automatically be made applicable. The somewhat dubious reasoning glosses over the fact that from the point of view of the jobs, and looking at the working environment, the comparison of an offshore installation and a merchant vessel is a superficial one. The background and training of the employees are those of totally dissimilar industries. The men join the industries for quite different reasons. The working conditions are not at all the same. The work on board an installation is quite different. The installation does not travel, is carefully controlled from the shore, and subject to different economic and social constraints. Although they might both be classified as 'total' institutions, an offshore installation is more of a productive unit and probably bears more comparison in sociological terms with an isolated factory. The installation is normally stationary. The hazards are not normally those of a ship on the high seas, and indeed if the installation is being moved or is moving it is normally commanded by a qualified seaman anyway. It seems equally arguable that an offshore installation bears strict resemblance to a ship *only* at the time of sinking, in which case it is likely either to be beyond the help of even a qualified mariner, or in such a situation that a well-drilled crew of any background could take care of it under the authority of the man they respect for his knowledge of offshore drilling equipment.

In point of fact the recommendations of the Tribunal found their way almost immediately into the licences via the instructions given by the Minister under the Model Clauses. Instruction No.4[53] reads:

> (1) Whenever such installation. . .is at anchor or affixed to or placed on the sea bed with a view to it being used in connection with searching or boring for petroleum there shall be on board a person appointed to be in sole command or charge (hereinafter called 'the person in charge') of the installation. . .and of all other persons on board who shall be in particular charged with the duty –
> (a) to keep himself informed of all matters affecting the safety of the installation. . .and the persons on board and to ensure that he is known to all other persons on board. . .

In effect, therefore, the 'Sea Gem' disaster and the Report led to the imposition of a formal command structure on the industry. Although this is an unprecedented move, there is no evidence that it has made any real difference to industrial practice, or that it was badly received. Indeed, it is difficult to see how it can be criticised other than in its attempt to follow the example of the merchant service in a slavish fashion without the remotest examination of the model chosen as ideal for seemingly superficial reasons.

Given this background it is not surprising that the Mineral Workings (Offshore Installations) Act 1971 should seek to reinforce the position. Sections 4 and 5 make provision for the appointment, duties and powers of a suitably qualified 'manager' of every offshore installation.[54] Section 4 (1) reads:

> Every offshore installation, so long as it is in waters to which this Act applies, shall be under the charge of a person appointed to be or to act as manager of the installation, and the owner of the installation shall appoint to be installation manager –

(a) a person who, to the best of the knowledge and belief of the owner, has the skills and competence suitable for the appointment, and
(b) another or others to act where necessary in place of the installation manager,

and shall inform the Secretary of State of any appointment under this subsection by giving notice in the prescribed form and containing the prescribed particulars.

By subsection (3) the owner may appoint under section 4 (1) (a) two or more persons to be managers in rotation, and those appointed under section 4 (1) (b) will be able to act in place of any of them. The difference between a manager and a person appointed to act in place of the manager where necessary reflects the request for designated 'second-in-command' made by the 'Sea Gem' report,[55] but is one that has always been present in the industry.[56]

Regulations made under this section[57] set out the forms for notification of the Secretary of State of an appointment, or termination of appointment under section 4 (4) in the event of the owner being satisfied that the manager does not have the requisite skills and competence, although no reasons have to be given, and it should be noted that the form of the notice to the Secretary of State concerning the manager's appointment requires that particulars of the appointee's 'qualifications, skills and experience which are considered to make him a suitable person for the appointment' are included.[58] The regulations also provide that the manager 'shall ensure that a notice of his appointment is kept posted on board the installation in such a position that it can be conveniently read' on pain of a £50 fine on summary conviction.[59]

It is difficult to say how far the Act will have made any real changes in the industry. Certainly the man in charge will be more clearly defined, but it is inconceivable that a company should put an installation worth many millions of pounds out in the middle of the North Sea without a person being recognisably in control. As already explained, this was never the case. All the Act will have achieved, therefore, is the notification of the Secretary of State of the name of the person in charge and the enforcement of certain, as yet undefined, standards of management. But it might help in marginal cases where it is necessary for legal purposes to have sound evidence (which would have to be disproved rather than proved, as was the situation before the

Act) that a certain named person was in control.

In fact it is noticeable that the Act goes out of its way to ensure that there is a designated person in command *at all times.* This is achieved by section 4 (5), which places the owner under a duty: '. . .in order to ensure that an installation manager appointed under subsection (1) (a) above is on the installation when it is manned, from time to time to place a person so appointed on the installation, and to ensure that he remains there until relieved, or so long as it is manned'. If he fails to comply, the owner faces a £400 fine under section 4 (6),[60] as does the installation manager himself if, contrary to section 5 (1) he is: '. . .absent from the installation at any time when it is manned, except in case of sudden sickness or other cause beyond his control, or for other sufficient reason. . .'.

The other encroachment into the area of traditional practice is in the provision for demanding qualifications of a person appointed installation manager. Section 4 (2) gives the Secretary of State power to make regulations, if he thinks fit:

> . . .prescribing requirements to be fulfilled as respects an installation manager. . .including requirements as to qualifications, experience, health or age; and the regulations may make different provision for managers of different types of installation or managers whose responsibilities differ in other respects and different provision for managers appointed under. . .paragraphs (a) to (b) [of subsection].

The Department has not as yet published any regulations under this section, and the evidence is that it does not intend to do so in the near future. The main obstacle has been that those in command of installations have no common qualifications apart from experience, and their experiences are unlikely to have been either uniform or of the same kind of installations.[61] The situation was summed up by the then Under-Secretary for Trade and Industry when he said in the House of Commons:

> . . .they have evolved as very responsible and excellent people. Nobody is criticising them. But they do not in any sense seem to have a common background of experience or knowledge which it would be possible to enshrine in regulations at this stage. I believe that it is right, for the short period ahead, to leave it to the operators and owners of rigs to choose the most suitable from their knowledge of those who have experience in this field. If in due

> course it becomes clear what the relevant qualifications are we have powers. . .to make qualification regulations. It might well have the effect of making regulations which rule out admirable people, or which allow in people, who, perhaps, do not have the requisite qualifications, if we were to attempt to do so at this imprecise stage of knowledge.[62]

It is submitted that very little change in the situation will take place in the foreseeable future. Indeed the Tribunal said of the qualifications for a manager: 'It would only be necessary for him to be a mature and responsible person trained to have a wholesome respect of the sea and knowing what to look for in carrying out his task of keeping everybody up to scratch.'[63] This seems to describe the present qualifications of toolpushers in charge, and explains why the person in that position will continue to fulfil the functions of manager.

2. *Other Crew Members*

The comprehensive safety regulations made under section 6 of the Act also lay down a number of personnel requirements. The manager must appoint a competent person as 'helicopter landing officer' to take charge of all helicopter operations. There must also be on each rig when manned at least one fully trained radiotelephone operator holding a certificate of competency for the rig's equipment issued under s.7(1) of the Wireless Telegraphy Act 1949. Thirdly, on every installation which is normally manned there must be at least one (two if there are more than 40 persons on board) medically trained person who is either a registered nurse or an enrolled nurse within the meaning of the Nurses Act 1957 or the Nurses (Scotland) Act 1951 or the holder of a certificate of competency issued within the last three years by either the St. John Ambulance Association, the St. Andrew's Ambulance Association or the British Red Cross Society, and who has received adequate training in the use of the obligatory mechanical artificial respiration equipment which must be carried on board.

Finally, a 'sufficient' number of 'competent persons' (undefined) must be on board appointed by the manager to be responsible for the control and safety of the structure of the installation, its electrical, mechanical and lifting equipment, drilling and production operations, the handling of dangerous substances, and any other unusual or dangerous operation. Ratification of the industrial structure by the law is thus complete.

4 THE TERMS AND CONDITIONS OF EMPLOYMENT

> 'The assumption that men seek to maximise their own return – to make as much money for themselves as possible – has an attractive unsentimental quality. It would be good were man so constituted that he would labor on behalf of others. Taking him as he is, this is not really to be expected and certainly not in modern business.' (J.K. Galbraith, *The New Industrial State,* Hamish Hamilton, London, 1967, p.115.)

It is commonly thought that oil-rig workers earn huge salaries. It is well known that the attitude of American corporations to trade unions differs from that of British companies, and one might be forgiven for thinking that working on an offshore oil-rig would mean participation in the excitement of new oil discoveries vital to the nation's economy. All these observations raise points which are relevant here, and highlight problems with legal significance related to the contract of employment.

In fact, many of the problems of the contracts of employment of offshore oil-rig workers are products of the structure of the companies for whom they work, the history of the industry and the pressures under which it operates, as well as the physical needs of offshore work. Thus the contents of the contract will depend largely on the recruiting policy of the company, the status of the worker, whether or not he is of British nationality, and whether or not the Contracts of Employment Act 1972 is applied. Some of these problems would be raised with respect to any contract of employment, but the range of options open to a large multi-national oil corporation operating throughout the world, employing workers in many different countries and expecting many of them to move from place to place, is much greater than any company, no matter how large, that operates only in Britain.

Some consideration must therefore be given to the various types of company involved, to the status of trade unions and to collective bargaining within the industry. The chapter also seeks to explain the form of the contracts and their background, to discuss the express terms such as wages, hours and holidays, and the various implied terms which bear on the issue of secrecy and confidential information.

I. CORPORATE STRUCTURE

Corporate structure and behaviour are obviously important in determining employment policies and industrial relations attitudes. Although the development of the oil industry has involved a large number of companies of all kinds, not all of which are oil companies in the traditional sense,[1] the basic concern in this work is with those companies that take on one or more of four separate roles: concession holder, the licensee of a specific area of the continental shelf for the purposes of either exploration or production; operator or field developer; drillers who perform the actual operation and generally own the rigs being used; and sub-contractors of one kind or another who provide ancillary services.

Concession holders range from merchant banks and investment companies to the huge multi-national oil companies. They include other corporations seeking to diversify, such as ICI. Many are the subsidiaries of large companies whose parents are almost unknown in Britain. The licensees of one of the blocks covering the 'Thistle' field for example (the 'Halibut' group) are Burmah Oil, Champlin Petroleum Co. (a subsidiary of the Union Pacific Company), Sante Fe (UK) Ltd (a subsidiary of Santa Fe International Co.), Deminex (a subsidiary of Deutsche Erdölversorgungsgesellschaft), Tricentrol North Sea Ltd (a subsidiary of Tricentrol Ltd), and Charterhouse Securities. Conoco (Continental Oil), Gulf Oil and the BNOC are the licensees of the other block. Licensees of the 'Ninian' field include BP, ICI, National Carbonising Co. Ltd, and subsidiaries of American oil companies such as Chevron (Standard Oil of California), Murphy and Ranger Oil. Other oil companies hold successful licences, including Shell, Esso (Exxon), Texaco, Occidental, Getty, Mobil, and Amerada Hess, but so do Rio Tinto Zinc and Allied Chemicals.[2]

Sometimes one or other of the licensees will be a company able and willing to do the development work itself, and will itself be the operator, such as Shell. In the case of the Halibut group an independent operator, Oil Development, was appointed. Some of the oil companies, such as Shell and BP, do their own drilling. Others may not, and they may in turn contract with drilling companies or companies that own rigs and provide drilling services such as the South-Eastern Drilling Co. (SEDCO), the Offshore Drilling and Exploration Co. (ODECO), Santa Fe, Global Marine and Transworld, to do the drilling for them. Often even those multi-national vertically integrated companies that are explorers, producers, transporters and distributors will turn to specialist drilling companies from time to time. The expense of drilling is such that capital tied up in rigs must be put to constant use. Even if the company has a number of rigs, additional fluctuating demand is met by the specialists who operate continuously on an international basis.

Any company operating or drilling in the North Sea has to depend on a variety of sub-contractors of one kind or another. There are specialist firms providing catering services on rigs, such as Chalk Drillcater; firms which supply technical equipment; seismic information; radio communications, such as Marconi; as well as helicopter services, such as Bristows, and supply boats bringing parts and pipes. The way such companies and the licensees, operators and drillers interrelate is of great importance from the employment point of view, and the pattern of sub-delegation has enormous repercussions for labour relations. Indeed, it can be important where safety is concerned. For example, in April 1975 complexities in company contractual and operating arrangements were blamed for the delay between the start of the search for the missing oil survey ship *Compass Rose III* and her disappearance.[3] It is not uncommon for several different companies to be employing different sections of the workforce on any one rig.

Whether or not the company is international in operations, or of United Kingdom or other nationality in origin, also has considerable significance in this respect. Looking at the companies in this way produces an entirely different picture. There is of course only one indigenous British company which operates throughout the world in this industry, British Petroleum, but it has played a key role in North Sea oil and gas discoveries. It has rigs of its own and for the most part does its own drilling and servicing. Sixty per cent of the Shell parent company is owned in the Netherlands, but this company has very firm roots in the UK and operates in a very similar fashion to BP. Other European companies involved include Total and Elf. But of course the vast majority of the operators and drilling companies are inevitably of American parentage,[4] and have inherited their style and attitudes from this background in addition to being heavily staffed from the USA.

Several consequences also flow from the fact that the industry is perhaps the most capital intensive in the world. For all the investment required in the North Sea, there are no more than 4,000 or so men actually working on board offshore installations, and few are indispensable. The financial background puts a huge amount of pressure on their activities. In addition, the North Sea provides its own deadlines. Most drilling has to be done in the summer months and only a few rigs can survive the winter in northern waters. The 'weather window' is even smaller for positioning production platforms, and it is here that missed opportunities or bad weather alone can set back the development of an oil field by a whole year – at enormous cost. The natural result is that delays, of any kind, are disproportionately expensive, and cannot be tolerated under any circumstances if avoidable. Here nationality is important, and a certain defiance of native attitudes is sometimes encountered in the removal of difficulties. Once the rig is in position, at an operating cost of over £100,000 a day, when the cost of support ships and derrick barges are included,[5] labour problems, which unlike geological difficulties are capable of some kind of immediate solution, are out of the question. Every care is taken, therefore, to ensure the loyalty and co-operation of the employees. If this is not forthcoming they are dispensed with, and replacements sought.

A similar attitude is to be found in the readiness of foreign companies to buy out (sometimes at a high price) the restrictive practices of the older surrounding industries which might continue to be tolerated by British concerns. A good example is provided by the difficulties presented by dockside stevedoring practices. Conoco have used a varying number of boats to serve their rigs off the coast of East Anglia. The normal number is two boats to one rig. But the number of rigs has fluctuated, and dockers at Great Yarmouth refused to allow boats to be unloaded by the same gang at the same time. One gang had to be paid for four hours at a time, even if the job only took two hours, with the consequence that one boat had to stand idle or the company had to have a further gang to unload the other boat, and in either case pay for unworked time. Other restrictions included paying for the equivalent of two extra men per gang when the load was 'dirty'. A typically American solution was sought. The company now keeps a gang hired by the week. Even in 1973 it was costing £500 a week, but the work is done as and when the company directs. Other examples can be found, but the effect of such attitudes on industrial relations, and on the adherence to legal requirements, can be readily appreciated.

II. COLLECTIVE BARGAINING

No study in labour law seeking to document the terms and conditions of workers in a particular industry can ignore the role and force of collective bargaining. Indeed, the most important elements of the contract of employment are usually found in collective agreements made between employers and trade unions. The individual contract has a decidedly inferior status as a bargain or as a document.

This is not so in the case of offshore oil-rig workers. The reasons are simple and straightforward: there is no official trade union activity on board offshore installations, no trade union recognition, and no collective bargaining. Although there are signs of change in this situation, some explanation is necessary.

The principal features of the offshore industry for the purposes of unionism are several, and indeed when they are put together it would not be hard to compare it with a modernised version of the situation faced by shipping unions almost a hundred years ago. It is a young industry feeling its way in a period of rapid technological change and internal growth. It is stretched financially and under enormous political and financial pressure. The work itself is hard, dangerous and frequently dirty, demanding separation from home, and thus performed by tough individualists frequently on a casual basis. The industry is one of the most capital intensive of all and the labour pool, subject to fluctuations, has often been large. In addition, most of the American companies operating offshore have imported their own brand of industrial relations and general hostility to the aims of British labour unions. Although there is evidence that their attitudes are changing with experience, many of these features would alone make unionisation difficult. Together they have made it almost impossible.

However, it would not be true to say that there are no union members at all out on the rigs. Many radio operators are members of the Radio and Electronic Officers' Union. Some of the barge engineers or mariners will be members of the Merchant Navy and Airline Officers' Association or the National Union of Seamen, and these unions have solid membership on the supply boats, indeed 100 per cent NUS membership among ratings. Some fitters and welders will be members of the Boilermakers' Union or the AUEW. There are electricians in the EPTU and technicians in APEX. Similarly, some of the roustabouts will have been in the TGWU prior to working on the rigs and will have remained in membership. The same is true of catering staff transferred from work on shore.

But although tolerated by some companies such as BP and Shell, union membership is actively disliked by others. An official of the Santa Fe oil company has been publicly reported as saying that the company went through 18,000 British workers to find the right 350.[6] There is a belief that unions and oil don't mix: the expense of running a rig trying to beat the weather means that work cannot stop for talks about petty grievances. Orders are carried out immediately or dismissal follows. The recruitment preferences of the companies, which include ex-servicemen and ex-seamen, militate against unionisation, as does the way in which the British worker, often a 'loner' in any event, applies for the job, pulled by the prospects of hard fast money, frequently from the other side of the country. In addition many workers are foreigners, often Americans themselves, and the high proportion of skilled workers, the benefits given to long-service employees, and the 'career plans' of the professionals (to whom the skilled workers are very close) are likely to produce company identification and the idea that unions are irrelevant.

But the various unions have all been keen to see an increase in offshore membership, particularly in view of the treatment some of the workers have received. In early 1972 and 1973 the TGWU and the AUEW made strenuous efforts to recruit offshore oil-rig workers but made little headway.[7] Indeed there is evidence that attempts by individual union members on some rigs to put up union notices or hold meetings have met with hostility from both management and other workers, and a number of incidents of labour troubles have been reported. In December 1973 some fifty British cooks engaged by the Norwegian firm Christiana Steam Kitchens for work on Phillips Petroleum oil-rigs were given thirty minutes' notice, and replaced by Spanish workers at lower rates of pay.[8] On 12 April 1974, twenty-seven members of the crew of the rig 'SEDCO K', being operated by Santa Fe for BP off Caithness, were all flown ashore for a 'cooling-off' period following a dispute on the rig over hours of work, the men demanding that their hours be shortened to the two weeks on/two weeks off being operated on other rigs in the area at the time.[9] The men were immediately replaced by workers trained at the SEDCO training school. As many of the replacements had been there for only a very short period of time and had not completed the course, the incident has led to the school being viewed with a degree of suspicion by unionists as a reserve labour pool for use in emergency. More recently thirty-five men working for an Aberdeen company on the construction of production platform 'B' for Shell's 'Brent' field were brought ashore after a dispute following their being asked to work in icy conditions 100 miles off the Shetlands in January 1976.[10]

In an attempt to counteract these and other similar practices, which do not seem to be confined to the British sector,[11] and in particular because of concern over the safety record of the companies and the apparent lack of any proper disciplinary procedures, an inter-union committee was formed in October 1972 by shop stewards in the shipbuilding yards in Aberdeen. Known as the North Sea Oil Action Committee, it is now parallelled by a committee called the Inter-Union Offshore Oil Committee (the IUOOC), formed in April 1974 from full-time officials of the TGWU, the AUEW, the NUS, the Boilermakers', ASTMS, APEX, UCATT and the ETU. In December 1974 a further separate association, the Association of Professional Divers, was formed to look after the special interests of these highly skilled workers who run arguably greater risks in the North Sea than any other group.

The reaction of the companies to the unions, however, has essentially been one of non-co-operation. In May 1974 the IUOOC invited representatives from twelve companies to a meeting, but all refused. In fact BP, Shell/Esso, SEDCO and a small company called Storm Brill agreed to talks with the unions,[12] and although initially unwilling to let union representatives onto the rigs themselves, a number of companies agreed to this by the end of the year. On the other hand, a recruiting drive by the unions at Dyce Airport, Aberdeen, when the crews were about to fly out to the rigs, failed to meet with much response, and a similar lack of success was reported by a TGWU official allowed onto the Shell rig 'Staflo' in early 1975.[13]

This confirms the conclusion that by controlling the type of recruit into the industry the companies have ensured that the individuals themselves are antipathetic to unionisation. In addition it must be remembered that some of the workers are recruited from distant areas and receive rail passes home. After a period of intensive work and isolated relaxation, the evidence is that the one desire of the men, a large proportion of whom have wives and families, is to get home as rapidly as possible. They have little incentive to approach unions if employed at high rates of pay on rigs providing excellent facilities, which is not always the case. The unions are, however, continuing to press their claims for recognition, and have indicated that their power to stop services to the rigs might be used to this end. This tactic has in fact already been used. A boycott was put on supplies and services to eight of ODECO's rigs in June 1975 over the dismissal of a union member after his demotion from crane driver to roustabout on the 'Ocean Victory'. The company said they would put him on another rig.[14]

The lack of any formal representation of offshore workers has had several results. In particular, section 7 (1) of the Mineral Workings (Offshore Installations) Act imposes the requirement that: 'Before making regulations under this Act the Secretary of State shall consult with organisations in the United Kingdom appearing to him to be representative of those persons who will be affected by the regulations.'[15] The regulations in fact cover topics generally thought to be the subject of proper trade union concern such as hours of work, training requirements, minimum age for employees, and health care.[16] In the House of Lords Debates on the Bill, Earl

Ferrers, the Government spokesman, said of the Secretary of State's duty: 'He would discuss with the appropriate bodies – the rig owners, the licensees, the T.U.C. and indeed Lloyd's shipping people – precisely the best way to frame the regulations. . .'.[17] Although this statement reflects the Government's concern with safety matters, without some kind of recognition there is no guarantee that those actually employed offshore will be able to express any opinions.

This situation is in point of fact ameliorated in a number of ways. Firstly, the TUC were asked for their comments on the proposals for those regulations which have been published so far, and both it and the Scottish TUC support the IUOOC which has had at least one meeting with the Secretary of State for Energy. Secondly, some rigs, though not very many, have weekly safety meetings at which relevant issues are discussed. These may be used to transmit the feelings of the crews and any points they bring up. Thirdly, Department of Employment officials in Scotland have been talking to the various companies over a period of time, trying to indicate what British firms would be expected to do in the circumstances. But more important changes can, of course, be expected now that an Order in Council has been made under s.127 of the Employment Protection Act, extending the application of that Act, the Trade Union and Labour Relations Act 1974, and the Wages Councils Act 1959. Although it should be remembered that the Order does not cover those provisions of the Act not in force on 21 June 1976,[18] and another Order can therefore be expected, assuming that the provisions are extended to the fullest possible extent they will enormously advance trade unionism and collective bargaining in relation to employment on the continental shelf.

As far as individual union membership is concerned the provisions of sections 53 to 56 of the Employment Protection Act will give individuals who have action taken against them short of dismissal for the purposes of deterring them from being members of a trade union, or from seeking to join a union, the right to make a complaint to an industrial tribunal which may result in an award of compensation.[19] A similar right will be given to employees dismissed as a result of union membership, or as a result of their proposing to become members of a trade union, under the provisions of the Trade Union and Labour Relations Act 1974 relating to unfair dismissal.[20]

The Employment Protection Act also gives those employees who are or who become trade union members the right not to be prevented or deterred from taking part in the activities of the union (assuming it to be an independent union within the meaning of the

Act[21]) at an appropriate time, or penalised for doing so.[22] In fact this right is modified by the requirement that if the activities are to take place on the employer's premises the union must be a 'specified' one,[23] i.e. it must be named under a union/membership agreement[24]. or accepted by parties to the agreement as being the equivalent of a union which is.[25] However, a union may be treated as 'specified' if it has referred to the Advisory, Conciliation and Arbitration Service a recognition issue which covers the employee in question, or if the Service has made a recommendation that the union covering the employee be recognised.[26]

Indeed it is the provisions of the Employment Protection Act relating to the recognition of trade unions which may turn out to be crucial for unionism on the continental shelf given the attitudes of many of the employers concerned. Under section 11 of the Act a union may refer to the Service an issue arising from a request for recognition by an employer, and this includes a request for further recognition.[27] The Service is then bound to examine the issue and consult the parties involved,[28] and may ultimately make a recommendation for recognition.[29] In the event of the employer refusing to take any notice of the recommendation, the union can then make a further claim to the Central Arbitration Committee, setting out the terms and conditions which the union thinks should be included in the contracts of employment of the employees covered by the recommendation. The Committee, after it has heard the complaint, and if it finds the complaint wholly or partly well-founded, may then make an award which may in fact be exactly what the union demanded. This award then takes effect as part of the contracts of employment of the employees involved,[30] and it is expressly provided that for the purposes of the Act it is immaterial whether the law which governs any person's contract of employment is that of the United Kingdom or any part of it.[31] These provisions are thus a very powerful incentive for employers to come to terms with trade unions and to bargain with them over terms and conditions of employment if the ACAS thinks this should be done, and the fact that the Service was looking at the issue of representation on offshore rigs even before the Act was passed gives some idea of the degree of importance being attached to the issue.

Union recognition is also relevant in a number of other important ways. Although the relevant sections are not yet in force, when they are, if a recommendation for recognition is made by the Service, or if a union is recognised by an employer, the union is entitled to receive a variety of information from the employer to help it in the collective bargaining process,[32] and again the sanctions for failure to produce the information are potentially considerable.[33] The exact nature of the information in question is not clear at present, but future determinations by the Service may eventually give a much better indication of what kind of information is not considered likely to cause 'substantial injury to the employer's undertaking for reasons other than its effect on collective bargaining'.[34] Union recognition would also bring with it the right of an employee who is an official of the union to have time off with pay for carrying out trade union duties,[35] and indeed to take part in various trade union activities.[36] It should be noted in this respect that although these individual rights are dependent on a further Order being made under section 127, as they were not covered in the first, those sections relating to the recognition which are in force may even have been operative without. As the trade unions and the companies operate on shore, where the bargaining would also take place, and as awards of the Central Arbitration Committee take effect through the contract of employment, if the contract was British the only problem appeared to be the authority of the Service outside the limits of territorial waters. But consultations and ballotting[37] could take place ashore. This issue has now, of course, been superceded.

The recognition of trade unions by most of the employers operating rigs on the continental shelf would have one further consequence for those employers who either chose not to do so, or were not made the subject of a referral to the Service even though they did not recognise the unions in question, or for new independent employers moving into the North Sea. Under the provisions of Schedule 11 to the Act, which, when in force will replace section 8 of the Terms and Conditions of Employment Act 1959,[38] a claim might be reported to the Service that as respects any worker an employer is observing terms and conditions of employment less favourable than those which are 'recognised', i.e. settled by an agreement between employers' associations and trade unions representing a substantial proportion of the employers and workers in the industry. If the claim is upheld, the Central Arbitration Committee may make an award in respect of those workers which takes effect through their contracts of employment.[39] Such a claim may only be presented by either an employers' association[40] or a trade union party to the agreement.[41] However, where there are no 'recognised' terms and conditions of employment, a claim may also be presented in respect of the non-observance of the 'general level' of terms and conditions in the section of the industry or trade in which the employer is engaged[42] by either an employers' association having members engaged in the same section of the industry, or by a trade union of which the worker concerned is a member.[43] This latter provision might prove more important if an Order is delayed as section 98 of the Employment Protection Act is not excluded from operation in relation to those who work outside Great Britain, and claims might be brought in relation to contracts entered into in Britain. But if the employer in question does recognise a trade union, only that recognised trade union may bring the claim.[44]

Finally, in the event of the trade unions making little further headway in the offshore industry, and the ACAS not making any recommendations for recognition, the Secretary of State for Employment still has the option of using the extension of the Wages Councils Act 1959 to employment on the continental shelf as a first step to the establishment of collective machinery.[45] He may refer the question to the ACAS first,[46] but if a Wages Council were established, it would have power to set the remuneration, holidays and other terms and conditions of employment for all or any of the workers that it covers.[47]

As far as collective action is concerned, the extension of the Trade Union and Labour Relations Act 1974 has several important effects. But its amendment by the Act of 1976 is even more relevant, removing the limitation on the definition of a trade dispute in section 29 (3) of the Act relating to matters occurring outside Great Britain. As the section used to stand[48] a trade dispute only existed in relation to matters occurring out on the continental shelf if the person or persons taking action in Great Britain, said to be in contemplation or furtherance of a trade dispute, were likely to be affected by the outcome of the dispute in respect of those matters laid down in section 29 (1). As the principal weapon of the unions seeking representation on the rigs has been the sympathetic support of the National Union of Seamen, who have 100 per cent membership on the supply boats, the removal of this limitation now allows NUS members to take industrial action in support of the claims even though they have no direct interest in the outcome. The provisions of section 13 of the Act would then come into play, but it must be remembered that although the amendments also take care of commercial contracts,[49] merchant seamen are in any event under special liability in such circumstances because of the provisions of the Merchant Shipping Acts.[50] The situation thus still remains full of complications.

Nevertheless, when these provisions are looked at all together, it seems that employment on the continental shelf may see some considerable changes in the near future. The potential application of selective Acts of Parliament in the labour field could bring the offshore industry within the mainstream of British industrial relations and end a decade or more of operation almost entirely within a free market situation outside the scope of the protective legislation that has covered workers ashore.

In the meantime, however, the absence of collective bargaining means some disparity in the terms and conditions of employment, particularly in matters of pay, both between companies and in some cases between different rigs. Salaries and wages are not negotiated but set by the companies, and indeed in all matters there is therefore room for enormous disparity of treatment. In addition, in stark contrast to the related merchant shipping industry, it also means that information relating to wages and other terms and conditions is unreliable in many respects and frequently varies enormously. It is also susceptible to rumour and exaggeration.

III. THE CONTRACT OF EMPLOYMENT

The emphasis of the individual contract in labour law is partly of historical origin. But it is partly a reflection of the dominant assumption that law is a mechanism for regulating only individual disputes. The consensus model of society refutes in itself the possibility of deeper divisions of interest. This emphasis is normally challengeable even without recourse to an alternative model of society. The importance of the individual contract is reduced by collective bargaining – but not obscured by it in legal theory. At the same time the chances of any real individual bargaining on the terms and conditions of employment of a manual or semi-skilled worker in a large corporation are virtually non-existent. But it would be wrong to minimise the importance of the contract of employment for these reasons. Though workers covered by collective agreements may be in a majority, even they still have their individual contracts. Just because collective aspects of industrial relations dominate considerations of labour law and industrial reality, it would be wrong to ignore those who only have an individual contract of employment, and whose bargaining strength is only that of the free labour market.

Offshore oil-rig workers are such a group. Their collective weakness makes their individual terms and conditions of employment of great importance. As will be seen they can vary considerably. But it is not only the terms and conditions that lack uniformity. Company policy may also dictate different sorts of contracts, and bearing in mind the legal uncertainties discussed in Chapter 2, different companies may have varied in their attitude to the application of legislation, even if this possibility will almost certainly disappear in the near future.

It is therefore proposed to deal with the contract in two parts: the form, and then the terms.

A. The Form of the Contract

The basic distinction to be made between the contracts of employment of offshore oil-rig workers is one which often follows the division between professional and non-professional employees. This division is further complicated, however, by problems of nationality brought about by the international operations of the companies involved and their recruitment policies.

In the case of British Petroleum, for example, whose employees are mainly British, all the roustabouts and other semi-skilled workers below professional level, such as rigmen and crane drivers, were for many years on short-service contracts, i.e. contracts of indefinite duration but subject to minimum periods of notice on either side. The reason for this was historical: the company was unwilling to hold out the prospects of any sort of career in the early gas days when continued employment was uncertain. The professional staff, by contrast, were already employed by the company, and were moved from other overseas operations. Centrally recruited, they all had staff contracts. These generally provide for longer periods of notice by the company, and give the right to participate in the company's pension scheme. Although BP's standard contracts have now been changed, and even unskilled employees can join the pension scheme, differences remain, and the contracts of professionals omit references to obeying the orders of immediate superiors, and of course differ over various details relating to pay and discipline.

Shell have a similar system which reflects their more complicated company structure. Although nearly all those who work in the offshore industry have short-term contracts, the company have a universal standard form staff agreement for ex-patriates intended to cover those in the professional category. The contract will be stated to be governed by either English or Dutch law,[51] and this depends on which part of the organisation recruits them. Usually, employees are recruited by the local companies, but those of professional status will normally join the Dutch or English companies, and can then expect to be sent anywhere in the world. The situation is further complicated by the number of employees on loan to Shell (UK) Exploration and Production from related companies such as Deutsche Shell, and Shell NV.[52] These employees will have contracts governed by the law of the country of that company. The employee will then have a letter of assignment, and be informed of the additional conditions of the contract. The most important of these will be his agreeing to obey the borrowing company's orders.

A similar distinction between the professional employees and other workers is made by most of the American companies. In this case American personnel will almost certainly have contracts made in the USA which will be governed by the law of the state in which the parent company is incorporated. Almost all these workers will be in the higher grades. Those workers recruited in Britain to work for one of these companies will almost certainly have contracts which reflect British law.

Turning to the contracts themselves, particularly those of the non-professionals, most of them even now cover all the subjects laid down in section 4 of the Contracts of Employment Act 1972. In fact, a note of the terms and conditions is normally unnecessary as most employees receive a written contract at the time of engagement.[53] This is particularly interesting in view of section 12 of the Contracts of Employment Act which states that:

> (1) Sections 1 to 5 of this Act shall not apply in relation to employment during any period when the employee is engaged in work wholly or mainly outside Great Britain unless the employee ordinarily works in Great Britain and the work outside Great Britain is for the same employer.
> (2) Subject to the foregoing subsection, this Act shall apply whatever the law governing the contract between the employer and the employee.

Before the extension Order under the Employment Protection Act was made, the Act obviously did not apply to offshore oil-rig workers. Work on an offshore installation would appear to be 'wholly. . .outside Great Britain' if the workers only ever worked on the rig, thus excluding them from the Act. The problem arose with those workers who were sent out to the rig to perform tasks from time to time, such as marine officers, or company welders, or indeed, some of the professional staff. The Act did apply to all those whose work was not 'wholly or mainly outside Great Britain', and to those who worked on the rigs periodically, but who 'ordinarily' worked ashore for the same employer. The decision as to whether an employee fell into either of those categories would have been factual, but extremely difficult,[54] and was a point which might have had to be determined by a court or an industrial tribunal. Extension of the Act by the Order made under section 127 of the Employment Protection Act will obviously make the whole situation much less complicated.

The previous situation, however, was a good example of employees' rights being dependent not on the force of law where the law might not be applicable, but on the 'goodwill' of the employer. The view was expressed that the men should not be given rights to which they were not legally entitled, but this attitude did not prevail for the sake of administrative convenience. It would have been hard to sort out those entitled under the Act where the position in some cases depended on interpretation of the word 'ordinarily'. It was easier, and obviously better industrial relations from the point of view of the company, to give all the workers the rights of the more fortunate in law. In view of the complications, therefore, it is not surprising that when the contracts were made available to the author, most of which were issued in 1973, it was found that many of the companies complied with the Act whether it was strictly applicable or not. The contracts issued by ODECO, Santa Fe, and Transworld, for example, all of whom are drilling companies, expressly stated the periods of notice in accordance with the Act.[55] British Petroleum provided for one month's notice for all employees after probationary period of one month during which they undertook to give one week's notice unless entitled to dismiss summarily. After ten years' service the company gave notice in accordance with the Act.

This was not the case of some companies, however. Global Marine gave notice of one day after three months' service. Chalk Drillcater followed the Act, but gave one hour's notice during the first twenty-eight days. SEDCO gave one week's notice after thirteen weeks' employment, and one day's notice prior to that time. The contracts did not seem to envisage, or mention, any longer periods.

Apart from the periods of notice, used here to assess compliance with the Act, the other subjects mentioned in section 4 were generally covered. All the contracts mention the pay scales, the right to holidays and sick-pay (if any), and the number of hours to be worked. These will be discussed below. On the other hand, section 4 (2) of the Act[56] stipulates that the statement given by the employer to the employees should also specify a person to whom an employee can apply for the purpose of seeking redress of any grievance relating to his employment, the manner of application, and any steps consequent upon the application. But as this particular requirement was bound up with the application of the Industrial Relations Act at the time, it is not surprising that few contracts contained such a statement. But Shell indicated that they took note of the Industrial Relations Code of Practice.[57]

B. The Terms of the Contract

The details of the contracts discussed below relate mainly to the non-professional drilling crew. Professionals will usually have slightly different contracts, reflecting both their higher status and position, and also their increased individual bargaining power.

1. Hours of Work

The way in which the shift system works on board an offshore installation has already been mentioned. Most non-professionals concerned with drilling will work a twelve-hour day for either fourteen days (168 hours) or seven days (84 hours), followed by an identical period off. In some cases they may work additional hours if required. Semi-skilled workers such as welders may sometimes only work during the day. Those with Global Marine work a ten-hour day, five days a week. In some cases members of the catering crew may work even shorter hours.

Professional grades and administrative personnel will normally not work in a shift system. They will be expected to work as and when necessary, in some cases twenty-four hours a day if the drilling is at a critical place.

2. Rates of Pay

The remuneration of oil-rig workers is calculated in several different ways. These can be detailed as follows:

(a) Basic pay. A distinction in pay scales must be made between those workers who receive salaries, as they do, for example, with Shell and British Petroleum, and those who are paid by the hour. A further distinction is possible between the scales of professional and non-professional workers.

It is in fact the salaries of the professional workers, the petroleum engineers, toolpushers and drillers, that have contributed to the widely held opinion that oil-rig workers earn large sums of money. In their case it is true.[58] The salaries match the skills of the worker, and represent the companies' investment in senior specialist personnel. In the case of foreign professionals, including divers,[59] salaries are paid at the rate payable in the home country. Americans are paid the very much higher American rates. Thus pay structures can be very uneven, and in some cases can cause friction between personnel. American divers, for example, might be paid up to £2,000 a week compared to £100 per week for some of their British counterparts in the early days, but recent expansion of the need for divers with the increase in the number of production platforms has raised the pay of British divers to American levels, and divers' salaries are regarded as being a matter for personal negotiation. It must be remembered, however, that to earn salaries of up to £20,000 a year the diver may sometimes have to live for weeks at a time in a pressurised chamber to avoid wasting time on decompression.[60] The risks are enormous. Radio officers earn over £20 per day on a freelance basis, although companies with shipping connections, like BP and Shell, hire their radio officers from the marine wireless companies like Marconi with equipment as they do for ships.

Staff salaries will be subject to periodic review by the company, and will change with promotion. At the same time, tax will only be paid in this country by foreigners if they are resident here for the requisite period of time.

Turning to the non-professional grades, some companies, including both Shell and British Petroleum, pay fixed salaries rather than time-rates. In 1973 a Shell roustabout had a *basic* salary of £1,128 per annum.[61] This would increase with length of service. The top basic rate for a derrickman with the same company was £1,695 per annum.[62] Other companies, such as International Drilling, a subsidiary of the Offshore Company of Houston, Texas, also pay by the month,[63] and a steward with Chalk Drillcater was receiving a weekly wage, not related to hours, of £25 per week. A roustabout employed by Transworld was receiving a monthly wage of £108.89 calculated on the basis of a basic rate of 40.28p per hour. It is important to note, in considering these rates, that it is only the companies that pay monthly that pay for the week ashore. In all other cases the wages are for the weeks actually worked (and will normally be paid for those periods of time the employee is on the rig and expected to work even if the weather, for example, prevents it) and they have to last the period spent ashore.

Hourly rates varied from company to company. ODECO paid 48p an hour to roustabouts, 58p an hour for roughnecks, and about £1.20 an hour for drillers. The managers of the Global Marine ships set their own terms and rates as they run the ships out of an allocated budget. But a welder employed by the company would get about £1.20 per hour. On board the 'Glomar III' roustabouts and 'cook's sidesmen' or galley assistants received a basic rate of 30p an hour. A welder with ODECO, by contrast, would get a basic rate of 78p an hour, but would work a full shift.

All salaries and rates are subject to review by the company from time to time. In the absence of collective bargaining, pressure for increases usually comes directly from the men themselves, although some companies take note of the cost of living, and all are concerned to offer reasonably competitive terms. Some were increased when the shift system was changed from two weeks on/one week off, which was then standard, to one week on/one week off so that no loss of pay was entailed.

(b) Overtime. No hourly worker who works a tewelve-hour day, eighty-four hour week, is paid the basic rate for all of those hours. The overtime structures vary only slightly. Most companies, including Global Marine, ODECO, SEDCO and Transworld treat the first eight hours Monday to Friday as basic hours. Time and a half is paid for the next four hours, and for the twelve hours worked on Saturday. Any hours in excess of these, and hours worked on Sundays and public holidays are paid for at twice the basic rate. There are minor variations from this pattern. For example, Reading & Bates pay time and a half for the first four hours worked on Sundays. Santa Fe pay the basic rate for the first four hours on Saturdays, time and a half for the second four hours and double time after eight hours on Saturdays.

This meant that an ODECO roustabout was getting a basic wage of £53.76 a working week and a roughneck £64.96. A roustabout with Global Marine, however, was getting only £33.60 a working week. By contrast the ODECO driller received £268.00 for his two weeks' work. These sums, however, were the wages for three weeks because of the shift system in operation at the time. This brought the basic wages of ODECO employees down to £35.84 per week for roustabouts, £43.30 for roughnecks, and about £90 for drillers. The Global Marine roustabout's basic weekly wage was £22.40.

(c) Allowances. In fact these figures did not reflect gross earnings. In many companies the workers were paid an extra allowance, sometimes called a 'platform allowance' for each day they spent on the rig. These

too varied from company to company. In 1973 Transworld paid £2.25 per day, British Petroleum, Global Marine and SEDCO paid £3 per day. Shell paid a platform allowance of £2.70. ODECO by contrast did not pay allowances, thus justifying the high basic rate. These changes brought the fortnightly wage of Global Marine roustabouts up to £109.20 (£36.40 a week). In their case the allowance was paid as:

> . . .full compensation for any and all disabilities associated with employment in the drilling industry including, but not restricted to: shift work; offshore allowance; work allowance; waiting time; travelling time; clothing allowance; working in adverse weather conditions or high places or restricted areas or wet places; dirt allowance; no pay during time spent off duty or between work hours; working with cement and chemicals; and any other hazards or inconveniences of offshore drilling operations.

In a unionised industry ashore, all these items would be bargained for separately.

(d) Bonus payments. Some companies paid bonuses in addition to platform allowances, mainly as incentive payments to reduce turnover. A Shell roustabout received an extra £387 per annum as a 'shift allowance'. Transworld paid a bonus of £10 per month for each completed calendar month of service, payable only upon completion of six calendar months' service. Other companies such as Shell paid good service bonuses from time to time, generally every two years.

(e) Miscellaneous. Before making an overall assessment of wages in the industry it is necessary to mention several other marginal factors. Firstly, some companies give rail or travel allowances to and from home and the depot. Secondly, payments are sometimes made in lieu of holidays. Finally, the food and board provided on the rigs is free to the employees. It means that if he is single the worker will spend little on himself during his tour of duty. The married worker, of course, is in a very different position.

Overall there is some variation in 'the rate for the job' paid by the companies for work offshore. A Shell roustabout was getting about £40 per week before tax, a roughneck about £50. At Conoco they were getting about £35 and £45 respectively. BP were then paying annual salaries of £2,750 and £3,250 respectively. Recent reports suggest that these figures have not greatly increased over time,[64] and that wages are considerably higher in the Norwegian sector. Most of the companies are in contact through the UK Offshore Operators' Association and other similar organisations, and are aware of the salaries being paid by their competitors, although there is no real evidence of even informal agreements as to wage levels. The competitive nature of the industry has ensured that there are variations, particularly among newer companies trying to break into the North Sea.

3. Holidays

As most workers in the offshore industry have every alternate fortnight or every other week off, it is not surprising that many companies do not give holidays. In some companies, however, the workers can expect a fortnight's holiday – usually taken in lieu of a shift, making at least four weeks between work periods – or payment instead.

In 1973 the contracts of SEDCO, Santa Fe, and Global Marine all stipulated that the annual holiday entitlement of employees in the calendar year in which they joined the company was at the company's discretion. Thereafter, employees were entitled to two weeks' holiday in each calendar year, with pay during the holiday equivalent to the basic rate for a forty-hour week, provided that not more than one week's holiday could be taken by employees until they had completed six months' service with the company. Holidays had to be taken at a time or times convenient to the company, and could not be carried forward from one calendar year to the next. There was no entitlement to unpaid holidays, but these could be granted at the company's discretion. Payment was only made for public holidays if the worker was actually on the installation, in which case he normally got double time. Other companies worked on a similar basis, though some calculated a day's holiday for every fourteen days' work.

By contrast, Chalk Drillcater did not give holidays at all. Neither did ODECO, but they paid a holiday bonus equivalent to two weeks' basic pay for each year of service. The employee was only eligible after a full twelve months' employment, and in their case the bonus was calculated on the basis of eighty-four hours per week. Transworld paid an allowance in lieu of leave at the rate of about 6.8 per cent per month of the monthly basic rate in respect of each completed calendar month of service, payable at the end of twelve months, or earlier on the cessation of drilling at the employer's discretion. As a result the employee was entitled to no leave at all apart from the seven day rest periods. All these figures could, of course, vary from employee to employee. In particular, professional workers will have quite different holiday allowances.

4. Sickness and Injury

Very few companies pay for days in which the employee is absent through sickness or injury. Frequently the contract of employment stipulates that the company does not operate a sick pay scheme, and that if the employee is injured in the course of employment he can make a claim through the Department of Health and Social Security.

Those companies paying monthly salaries, however, frequently pay half pay or even basic pay for periods of absence sanctioned by a doctor's certificate. This is generally only after the probationary period is over, and may be limited to a certain number of weeks in any one year.

5. Pensions

Generally speaking few companies offer their hourly paid employees pensions of any sort. They are, however, required to contribute to the state graduated pension scheme, and all companies pay the National Insurance stamps. British Petroleum does offer its non-professional employees the right to join its pension scheme, and pays a gratuity on leaving equivalent to 5 per cent per annum for each year's service with the company.

6. Termination of Employment

The periods of notice given and required by the companies have already been discussed. In nearly all cases the contracts specify the acts for which an employee can expect to be dismissed summarily.[65]

7. Transport

Some companies inform the employee in the contract or in a letter that transport between the base on shore and the rig or ship may be by road, rail, boat, helicopter or fixed-wing aircraft. The employee is normally warned that he must be prepared to accept any method of transport designated by the company.

8. Safety and Equipment

In some cases employees are instructed to abide by the safety rules of the company, posted on the rig, at all times. In many cases companies provide protective clothing such as safety hat, boots and gloves and an oilskin 'slicker suit' at cost to the worker. In other companies they are issued by the company but must be returned if the worker leaves or is discharged.

9. Grievance Procedure

As already indicated, not all the contracts specified a grievance procedure in accordance with the provisions of the Contracts of Employment Act. ODECO indicated that a grievance should first be raised with the immediate supervisor either orally or in writing. If the matter was not settled in fourteen days, the grievance was to be put in writing and forwarded to the Drilling Superintendent. This sort of procedure is adopted by British Petroleum who stipulate in their letter of engagement for new staff that:

(a) When, during the course of your employment with the Company you have a problem or grievance relating to your employment, you should in the first instance discuss the matter with your immediate manager or supervisor.

(b) If you cannot resolve your problem with your immediate manager or supervisor, you have the right to require him to refer the matter either to the next senior level of line management, or to the appropriate member of your Personnel Unit.

By contrast, Chalk Drillcater told employees to raise their grievances with the UK Base Manager (Personnel) in writing, and if it was not settled, with a company director, also in writing. Most of these provisions will not have altered, and it should be remembered that nearly all the companies have their personnel departments or their staff centres on shore and if the problem is one that the installation manager finds is outside his responsibilities, such as pay, taxes, or holidays, then it will be passed on to the person in charge of operations on shore, or to the specialist department.

10. Confidentiality and the Duty of Fidelity

The duty of 'fidelity' is one of the more complex of the duties incidental to the employment relationship, and the details are best left to specialist works on the subject.[66] Nevertheless, because of its importance and the fact that it appears in the contracts, the main aspects relevant to the employment of offshore workers must be mentioned.

(a) Spare time work. Some oil companies, especially those with salaried

employees, restrict their employees from working for another concern during their off duty period. In other words they will extract the usual condition that the employee will devote his whole time to the company and not engage in any other business. This is certainly not true of all offshore operating concerns. To many, what the employee does in his week off, of his month off during which he has his fortnight's holiday, is entirely up to the individual employee. In some cases it is regarded as part of the attraction of offshore work that, in the example of a welder or a mechanic, the worker can have his own small business in which he works on his weeks off. Indeed the company would seem to rather expect that their employees would behave in this way, and the fact that 'moonlighting' in its various forms is more accepted in America than in Britain may well have something to do with this attitude. No doubt even these companies would draw the line, however, at the employee's working for a potential rival company, or in a position in which the employee might give away the company's secrets, even if he does not in fact do so.[67]

(b) Subsequent competition. Covenants to restrain the employee working for a rival after the termination of the contract with the company might well be used in the case of senior personnel in the industry. In fact there is no evidence of this, and employees do change companies, though there is certainly a degree of company loyalty amongst the trained specialists for whom the company will have 'career plans'. Nevertheless, provided the covenant were 'reasonable in the employer's interests', it would be permissible.[68] The fact that such a covenant, if drawn on a time basis, might have the effect of making it impossible for the specialist to pursue his skill would no doubt make it unreasonable and the covenant might have to stipulate that the employee must not work in a particular area, such as the North Sea, for another rival company. This seems incompatible with the way in which senior specialists are recruited, the sort of contracts they are given, and the way in which they frequently change their place of work. In fact, the company may well be adequately protected by the duty not to divulge confidential information, and in the oil industry it is hard to see how one specialist, under such a duty, could harm a company by going to work for a rival.

(c) Information of a secret or confidential nature. It is hard to say how much confidential information any particular offshore worker might receive in the course of his work, just as it is hard to draw a line

between information that would be considered confidential and that which would not. Nevertheless, given the nature of the employment, especially that of the specialists, the degree of competition, and the value to a rival company of information relating to an oil or gas discovery or to a seismological survey of a new region,[69] it is not surprising that many contracts for offshore workers do contain a clause placing the employee under a duty not to impart any information regarding the company's business or affairs at any time, whether during or after the contract, without the express permission of the company. The contract may well also contain a clause giving the company exclusive rights to any inventions or improvements in plant or machinery or in any processes discovered by the employee in the course of his employment. Obviously, the availability of any such information and the likelihood of making discoveries will be more common amongst some employees than others, but in any cases where an express clause is not found in the contract the courts would imply a term to the same effect.[70]

If an employee breaks the duty the company might seek an injunction against him, sue for damages for breach of contract, claim for an account of the profits gained through the breach, and in any event will probably be entitled to dismiss him. The company might also try to obtain an injunction against the person or company to whom the information was revealed,[71] even if that third party did not know that the information was confidential at the time it was passed to him,[72] and this may still be the case even if the information was innocently purchased in good faith.[73] The company might also obtain an injunction (though in all cases this remedy is discretionary) if the third party knew the information was confidential but used it without realising.[74] Damages might also be awarded when the third party knowingly induced the employee to divulge the secrets. He would be liable for inducement of breach of contract.

The most difficult problem is quite simply the classification of secret or confidential information. The cases seem to indicate a difficult distinction drawn between information which: '. . .can fairly be regarded as a separate part of the employee's stock of knowledge which a man of ordinary honesty and intelligence would recognise to be the property of his. . .employer and not his to do as he likes with. . .'[75] and that which is 'his to do as he likes with', usually regarded as embracing the 'individual skill and experience' or 'know-how' which the employee had acquired in his work.[76] It also seems clear that whereas written information is more susceptible of classification as confidential, the fact that the employee carries it around in his head need not necessarily make the slightest difference.[77] Furthermore, the information need not be very technical or detailed, it is primarily the value placed upon it by the employer that matters.[78]

In the case of oil-rig workers these distinctions are well illustrated. Vast amounts of technical information is available in trade journals, and even the details of oil-wells are often published.[79] As most offshore workers are largely 'on the job' trained, 'know-how' must form a great part of their information about the work. If experience plays such an important part in the training of a toolpusher or a good driller as the companies say it does, the area of confidential information about the activity is considerably reduced.

(d) Improvements and inventions. If the contract does not contain a clause concerning patents and inventions it would seem that any invention or improvement made by a worker belongs to the employing company and not to the individual employee. But the question turns very much on the status and job of the employee and the nature of the invention or improvement itself.[80] If it is totally unconnected with the company's affairs it might well be allowed to the employee.[81] In some cases apportionment may be ordered of the benefit.[82] If the material can be classed as copyright the situation is governed by the Copyright Act 1856, section 4, which has the effect of vesting the copyright in the employer if the work was done by an employee in the course of his employment.[83]

11. Other Clauses

Only clauses which are peculiar to the contracts of employment of offshore oil-rig workers have been covered here. In many cases there might well be exclusion clauses of some kind, especially in relation to accidents.[84] Other terms which are implied into the contract by the common law have been left to more general studies.[85]

IV. CLAIMS AND THE CONTRACT AND CONFLICTS OF LAWS

Oil-rig workers are not subject to special legislation relating to the payment of wages unless, of course, they are also merchant seamen employed on British drilling ships. For workers on ordinary offshore installations not so registered who enter into their contracts in this country, are resident here and are paid here, most of the ordinary law relating to contracts of employment, taxation, and recovery under the contract will apply.[86]

Although this would appear to make contractual issues quite simple, the presence of a large number of foreign workers employed by overseas companies on rigs on the British continental shelf in fact makes for a rather more complicated situation than might be imagined. As the continental shelf is outside Great Britain, foreign workers employed under foreign contracts from abroad will not come under English law. A classic example of what this means was provided by the application of the Counter-Inflation Act 1973, to the wages of offshore workers. Those companies who paid the wages in this country of persons working abroad were subject to orders from the Pay Board.[87] But French or American concerns not represented in this country were not susceptible to such orders and the wages of their employees paid abroad could therefore rise whilst those of their fellows employed by British subsidiaries could be restrained. A similar result would be achieved by the application of section 1 (1) of the Remuneration, Charges and Grants Act 1975 if wage increases were 'agreed' and then refused, as that Act would provide a defence in an action for breach of contract, but only if the law to be applied included that Act. On the other hand, were the provisions of the Employment Protection Act 1975 relating to guarantee payments to be extended to employment on the continental shelf,[88] it would be immaterial which law supposedly governed the contract of employment of the worker concerned[89] were he to bring an action before an industrial tribunal for payment in respect of any day on which he was on the rig but not able to work, for example because of the weather, and was not paid in consequence. It should always be borne in mind when considering such problems that the approach adopted by section 3 of the Continental Shelf Act 1964 is not to make the continental shelf part of Britain as far as legislation is concerned, with the result that although licensees have to be British, if legislation becomes too onerous it might be worth an employer's while employing people to work on the British shelf from outside Britain.

This situation means that a few words must also be said about problems of conflict of laws in relation to claims arising out of the contract of employment.[90] As this is a subject which is fully discussed elsewhere,[91] only the main principles touching terms of the contract already discussed will be treated here.

In the event of an action being brought in an English court on the contract of employment of an offshore worker, and that worker, or the employer, is foreign, the problem of the 'proper law' arises, summed up as 'which law applies to the proceedings?'.[92] In fact, the law that applies to one part of a contract may not be applied to every part,[93] though it remains true that it is only in rare cases that the court will 'split' a contract[94] and apply different laws to different parts. The English courts have, indeed, in their attempts to answer the question of which law applies, declined to adopt a rigid doctrine. They have refused to take any one of the factors that might connect a contract with a particular legal system as the sole criterion for identifying the law to be applied. But it must not be thought that there are no rules at all, indeed the very opposite is true, and they may be summarised as follows:

(1) Where the parties expressly stipulate that the law of a particular country shall govern the contract, the law of that country *generally* applies,[95] though the parties may incorporate a different system for any particular part of the contract.[96]

(2) Where the intention of the parties as to which law shall apply is not made expressly, the proper law is that which the parties ought to have intended, taking into account the circumstances of the case.[97] This 'reasonable man' test is sometimes reformulated as the application of the system which has the closest and most real connection with the contract.[98]

There are, therefore, two situations: in one the parties have made an express choice and the matter is relatively straightforward. The problem is to determine the law to be applied in the absence of choice. The courts have attempted to answer the question by considering a number of factors, none of which one could say was sufficient in itself. These include the residence and the country of origin of the parties,[99] the place where the employing company is incorporated and where its 'principal place of business' is located;[100] where the contract is made;[101] and perhaps even the language of the contract.[102]

In the case of contracts of employment the choice is most frequently between the law of the place of work and the law of the seat of the firm employing the worker, which is usually the same. In *South African Breweries* v. *King*[103] a worker of English origin living in Johannesburg was given a contract of employment in the English language with a company with its registered office, secretary and directors in England, but which carried on business only in South Africa. The contract was to be entirely performed in South Africa. The Court of Appeal held that the contract must be governed by the law of the Transvaal with which it had the closest connection.

By contrast, the law of the seat of the firm was held to apply in *Sayers* v. *International Drilling Co. N.V.*[104] In this case the plaintiff, an Englishman, contracted with the Dutch subsidiary of an American company to work on offshore rigs outside the United Kingdom. The standard form contract was drafted in Americanised English and was clearly intended for use in connection with workers of differing nationalities. His salary, however, was payable in sterling, and the company paid his National Insurance contributions.

The contract in fact purported to exclude the plaintiff's legal rights to claim against the company in the event of accident, and substituted a company 'Compensation Program'. Clause 8 of his contract stated:

> As the company is a Netherlands corporation, and as my employment contract hereby applied for will be wholly performable overseas and outside the United Kingdom, the company does not subscribe to or carry workmen's compensation insurance under the laws of the United Kingdom. Accordingly, I realise that I shall not be covered by virtue of my proposed employment with the company by workmen's compensation insurance or benefits under the law of the United Kingdom. . .I am satisfied with the provisions and benefits of the said Compensation Program.

and:

> In consideration of having this Program maintained for me, I hereby agree that, in the event I am accidentally injured or sustain disability of any kind during my employment with the company, I will accept those benefits to which I may be entitled under the Compensation Program as my exclusive remedy in lieu of any other claims, rights, demands, or actions whether at common law or under the statutes of United Kingdom or any other nation, which may accrue to me by virtue of such accidental injury.

If the plaintiff had been injured and the contract had been governed by English law, that clause would have been void under section 1 (3) of the Law Reform (Personal Injuries) Act 1948.

The plaintiff was sent to Nigeria to work as a derrickman/pumpman. Within a fortnight he was injured and flown home. Instead of being satisfied with the terms of the Compensation Program he sought to have the clause overridden by claiming that the proper law of the contract was English law.

The Court of Appeal dismissed his claim, and held, unanimously though for different reasons, that the proper law to be applied was Dutch, and therefore the clause was valid. Both Stamp and Salmon L.J. took the view that the contract had a closer connection with Holland. The company could change the position of the rig, so Nigerian law was dismissed. The words 'United Kingdom' in the clause were omitted in the original and filled in with the rest of the terms. Furthermore, as English law would have invalidated the clause, Salmon L.J. thought that this was a pointer to Dutch law. The parties may well have intended the contract to be governed by a system of law under which it would be valid.[105] On the other hand, Lord Denning M.R. thought that the proper law of the contract was English law.[106] In a more recent Scottish case a similar issue was raised in a claim brought on a contract governed by Norwegian law. Criticising Sayer's case, Lord Kissen held that where the law of the place where the cause of action arose and that of the place where the action was brought would both apply the Act, they should oust the law of the contract to render the defence void.[107]

It is therefore hard to lay down any set rule. It depends on the facts, and the emphasis given to them by the individual judge. The complexity of this situation was summed up by Professor Kahn-Freund in 1957 when he maintained:

> The need for flexibility in the criteria used for the determination of the proper law of the contract of employment arises from the generality of the concept of the contract of employment: it covers so many social phenomena that any formula of conflict of laws seeking to treat all cases alike would force an infinite variety of factual relations into too rigid a legal mould.[108]

The enormous variety of the categories of workers considered here bears out this conclusion. The work of many oil-rig workers such as Americans brought over to the North Sea, is intended to be wholly performed abroad, where the worker must take up residence. On the other hand the worker of 'professional' status is transient – his residence in that foreign country is temporary, and he might move to another foreign country rather than return to his country of origin. In such a case his home country might assume greater importance for the determination of the question of proper law, particularly if that is the country where his company has its head office. This distinction, made by Professor Kahn-Freund,[109] between higher and lower employees, between those who are employed as mobile and those who are employed to do a specific job, well suits the division in status between professionals and non-professionals common among offshore oil-rig workers.

5 SAFETY, HEALTH AND WELFARE

'If employers employ men on this dangerous work for their own profit they must take proper steps to protect them, even if they are expensive.' (Denning L.J. in *General Cleaning Contractors* v. *Christmas* [1952] 1 All E.R. 42 (C.A.).)

It has already been intimated that one of the causes of new recruits leaving the offshore industry might be the inherently dangerous nature of the work. Indeed, without direct experience it is hard to really appreciate just how dangerous it can be to work 200 miles off the Shetlands on an offshore platform lashed by gales in the winter months. Although increasing public and government concern is being focused on safety matters, and even the most basic economic considerations mean that the companies themselves are concerned that operations should be conducted without accidents and injuries, considerable physical courage is nevertheless required. Accidents do happen, and far more frequently than in almost any other industry. This chapter therefore looks at a number of interrelated topics relevant to the safety, health and welfare of persons employed on the continental shelf: the kinds of accidents that occur; the legislative steps that have been taken to reduce their number; and the measures taken to promote a healthy working environment and to minimise the hardships of life. The various legal consequences of accidents that do happen are dealt with in the following chapter.

I. ACCIDENTS AND INJURIES

Between 1955 and April 1974 there were at least sixty-eight major mobile rig disasters throughout the world, involving the loss of at least 100 lives and costing approximately $198 million. Some twenty-nine of the rigs, equivalent to about 12 per cent of the world total of operational offshore rigs, were total losses.[1] There have been others since.[2]

Not surprisingly, in view of the adverse weather conditions, several of these incidents have happened in the British sector of the North Sea. The 'Sea Gem' was lost forty-five miles off Grimsby in December 1965; and the 'Conoco I' collapsed and sank off Grimsby in April 1967. The 'Ocean Prince' was lost off Flamborough Head in March 1968, and in November 1969 the 'Constellation' was lost, again off Flamborough

Head, both while under tow. In January 1975 the 'Transocean III' sprang a leak in one of its supports and capsized 100 miles south of Shetland. The crew were flown off on to the 'Transworld 61' which then also ran into difficulties itself in the storm and nearly capsized as well.[3] Other rigs have been sunk in other sectors of the North Sea. For example, the 'DPI', a gas drilling platform, went to the bottom in 350 feet of water in the Norwegian sector in October 1974 when its metal flotation tanks burst. There have also been a number of very narrow escapes. In November 1972 fifty-three people were flown off the 'Ocean Tide' when it threatened to capsize in gale force winds while moving location.[4] A production platform for the 'Frigg' gas field was sunk some three kilometers out of place in February 1975 when its flotation equipment failed.[5] The weather has ripped other rigs off their anchorages. Mobil's loading tower for the 'Beryl' field broke away from its foundations in December 1975 and drifted towards the Norwegian coast.[6] Both 'Mr Cap' and the 'Glomar IV' were moved to Nigerian waters because they could not stand up to the conditions of the North Sea.[7]

Added to the weather, there are the hazards of oil and gas themselves. Pipes can rupture, gas or oil may ignite, and a field may 'blow-out', i.e. the pressure may force the oil or gas up the hole, destroying the well-head and the equipment at the top and causing considerable damage and injury. There were two gas blow-outs in 1968, and one on BP's 'Platform A' on 7 June 1971.[8] At the beginning of November 1975 a corroded oil pipe-line ruptured causing a massive explosion on board the Philips Petroleum production rig 'Alpha' on the 'Ekofisk' field which was then swept with flames.[9] Other major incidents have occurred as a result of collisions between supply boats and the rigs,[10] and in one case the boat sank, although the crew were all saved.[11] There has yet to be an incident such as that which occurred in the Gulf of Mexico in August 1975 when the tanker 'Globtik Sun', carrying 350,000 tons of crude oil, hit an unmanned oil platform off Texas.[12] But the combination of offshore oil and gas operations and bad weather is inherently dangerous anywhere, and when it is considered that conditions off the north-west of Scotland, and in the northern North Sea where oil has been discovered, are among the worst that have been encountered by offshore operators anywhere in the world, some idea of the enormous risks involved may be gained.

Although these dangers can perhaps be best appreciated only by experience, some figures make the situation reasonably clear: the percentage of waves over six feet in a measured period of time in the

Gulf is only 5 per cent, off the coast of Nigeria, 12 per cent, and in the North Sea, 29 per cent.[13] The first rig to commence drilling in the North Sea had to raise itself up from its position forty feet above the water level because it was being buffeted by waves.[14] Indeed, the 'Sea Gem' usually operated at a height of fifty feet above sea level.[15] Waves can in fact go to nearly 100 feet, with winds of up to 150 m.p.h. to match.[16] The temperatures reach freezing point with regularity. Furthermore, tidal motions can produce strong undertows, which cause dangerous instability in the case of 'jack-ups' by scouring the sea bed around the supporting legs. Drilling in the North Sea, in waters of up to 600 feet in depth, is pushing the industry's technology to the limit.

The number of personal accidents that have resulted from work in these difficult conditions is hard to estimate. All licensees of designated areas are under a contractual obligation to report to the Secretary of State any accident which results in death or a broken limb, or requires transport ashore for medical aid. But no provision is made for the publication of any figures, and of course serious accidents can and do happen which fall outside these definitions. But regulations made under the Mineral Workings (Offshore Installations) Act 1971, also require the installation manager to enter details of any accident in a log book,[17] and this in turn has to be given to the Secretary of State.[18] Indeed, separate regulations also require the notification of any 'casualty'[19] (defined as a casualty or accident involving loss of life or danger to life) and a return in relation to every accident, injury or disease suffered by a person on or working from an installation or attendant vessel in connection with the installation as a result of which the injured person is disabled from work for a continuous period of three days.[20] But again no express provision is made for publication.

Nevertheless some figures have come to light, although it cannot be certain that the Department of Energy's register covers them all, particularly those which happened before 1974. The official figures, which include the deaths of divers, show that for the years 1965 to 1974 there were forty-three fatal accidents to workers on installations and attendant vessels in the British sector of the continental shelf. Some were killed by falls from the derrick tower, others by whiplashing chains or hawsers or by moving pipes. Others drowned.[21] Thirteen resulted from the loss of the 'Sea Gem' in 1965. Of the twelve fatalities which occurred in 1974, six happened on installations and three on vessels engaged in related work. The remaining three were diving accidents. Although two more divers are reported in the official figures as having died from accidents occurring during pipe-laying operations, casualties associated with work on or from pipe-laying barges are not included in the figures, which also show that for the same period there were apparently some 175 'serious' accidents, 25 of them in 1975 alone.[22]

On the installation itself the workers most seriously at risk as far as injury is concerned are those who work on the drilling floor – the roustabouts and floormen working in filthy, cold, muddy conditions for long periods of time. But even they are not so much at risk as are the divers. Here there is room for argument over the statistics. According to information in the possession of the North Sea Oil Action Committee, up to the beginning of 1976 some twenty-seven divers had died in the North Sea since operations began. But other figures indicate that as many as twenty-four had died between the beginning of 1971 and mid-September 1975 in the British sector alone, including nine in 1974 and eight in the first nine months of 1975.[23] The official figure of three fatalities for 270 divers employed in 1974 produced a fatality rate of 11.1 per thousand employed, as against 1.6 for those employed on the rigs themselves. The higher figure would give divers a fatality rate some eight times that of deep-sea trawlermen, normally considered the most endangered category of worker, thirty-three times that of coal miners, approximately 220 times that of ordinary factory workers, and all in all at least four or five times that of the workers on the rigs themselves.

It is not surprising that these figures have given rise to considerable disquiet. In January 1975 the Scottish Council of the British Medical Association set up a twelve man working party to study the medical difficulties of oil-related injuries. One of its tasks was to examine the dangers of diving. In its report, the committee hoped that a special surgical team could be set up, based in Aberdeen, which would be able to fly out to treat an injured diver in a decompression chamber.[24] But a glance at some of the accidents themselves shows that in some cases the fatalities occur in situations where very little can be done in any event, or in a totally unforeseen manner. A good example of the latter type of fatality, well illustrating the fact that the divers are operating at the limits of modern underwater and medical technology, was provided in September 1975 when two experienced, competent divers working for Oceaneering International Services died in a decompression chamber aboard the drilling rig 'Waage II' from what was subsequently discovered to have been hyperthermia, i.e. overheating, a problem which no one had ever encountered before.[25]

In fact the more realistic causes of concern about diving have been two. Firstly, the rapid increase in the numbers of divers needed has brought many thoroughly untrained men into the industry. For years there has been no real standard tests for deep diving work, and the best-trained men, lured from the Royal Navy by the high wages as fast as the service could release them, have been in short supply. The temptation to expand diving services with divers whose experience and credentials have not been thoroughly proven has consequently been hard to resist. It is to counteract such tendencies that the diving school has been opened on Loch Linnhe, and the Training Services Agency has begun to monitor standards. The second cause for concern has been that until recently the only manual of safe diving practices has been voluntary,[26] while many of the sub-contracting firms employing divers have been operating under the considerable pressure inherent in an industry where time is at a premium and delays are enormously costly. This pressure is of course transmitted to the divers themselves, and the financial incentives naturally increase the willingness to take risks.[27] It takes three days of decompression to bring a diver back to normal atmospheric pressure after only one hour at 600 feet, but it takes the same time no matter how long the diver operates at that depth or at the pressure to be found at that depth. Some divers have therefore been asked to spend as long as three weeks at the same pressure, and to live in a pressure chamber for that time, to avoid the wastefulness of a disproportionately long period of decompression after each dive. Such practices explain the high financial rewards, but the complete physical effect on the diver concerned is as yet unknown.

Finally, as if to emphasise the extreme vulnerability of any fixed or floating platform in the North Sea, there has also been increased concern for the safety of offshore installations and the possibility of political sabotage. In August 1975 a warning was received that a bomb had been placed on the Philips platform 'Arpet Alpha' on the 'Hewett' gas field. The rig was evacuated, the gas shut.off, and Royal Navy divers and bomb disposal experts searched the rig but found nothing.[28] Even before this incident, however, the Government had decided to increase the strength of the special Royal Navy protection force keeping an eye on offshore rigs.[29] Five new 1,250 ton patrol craft are being built to take on this task, and eventually the force will consist of these vessels operated by the Navy and three more operated by the Department of Agriculture and Fisheries for Scotland, as well as a flight of four Nimrod maritime reconnaissance aircraft. These efforts are being co-ordinated with those of other countries,[30] and Norway proposes to spend some £180 million expanding her oil-rig and fishing vessel protection fleet to seven surveillance vessels, six helicopters, three reconnaissance aircraft as well as a special firefighting and rescue craft.[31] This latter aspect is considered so important by the industry that the UK Offshore Operators' Association is itself planning the construction of four special firefighting vessels for service in the North Sea. Although there are even now special units that fit on the deck of a supply boat, the Association has in mind special semi-submersible craft capable of staying on the job no matter what the weather conditions are like.

II. PROTECTIVE LEGISLATION

With this kind of background it is not surprising that safety has been a focal point of government interest. Although naturally aware of the extreme conditions and the potential hazards of the North Sea, nothing brought home the point more forcefully than the collapse of the 'Sea Gem' in late 1965, not long after operations had begun, and it can fairly be said that the highly regarded Report of the inquiry which followed provided considerable impetus for the legislation making provision for safety regulations. However, although most of the regulations under the Mineral Workings (Offshore Installations) Act have now been made, the Act itself is not the only legislation bearing on safety on offshore operations. Several Acts impinge upon activities on the continental shelf in a number of ways, and it is proposed to deal with them in turn.

A. The Continental Shelf Act 1964 and Instructions to Licensees

As the first piece of legislation covering activities on the continental shelf it is not surprising, given that it was primarily concerned to create a legal framework so that drilling could commence, that much like the initial legislation passed by other countries around the North Sea the safety provisions included in the Continental Shelf Act 1964 were somewhat rudimentary. Apart from the application of certain provisions of the Coast Protection Act 1949 requiring permission to be granted where works on the continental shelf in designated areas might be an obstruction or danger to navigation,[32] which in any event would not apply to works carried out under a licence issued under any other Act,[33] these provisions were of two kinds, those relating to safety zones, and, of more importance, those relating to model clauses in licences issued under the Petroleum (Production) Act 1934.

1. Safety Zones

In line with the International Convention and the legislation of most of the other countries around the North Sea, the Continental Shelf Act makes provision in section 2 for the creation of safety zones around installations. These have the effect of prohibiting ships from coming within 500 metres of an installation without permission from the Department. A number of Orders designating such zones have now been made.[34]

2. Instructions under the Model Clauses in Licences

As already described the earliest method of government intervention in safety matters, still in operation, was through the model clauses incorporated in the licences granted under the Petroleum (Production) Act 1934, by virtue of section 2 (2), (3) and (4), and section 6 of that Act, and section 1 (3) and (4) of the Continental Shelf Act 1964. Under the regulations[35] the licensee is bound to comply with any instructions given by the Minister in writing for securing the safety, health and welfare of persons employed in or about the licenced area. The unsatisfactory nature of the remedies available in cases of non-compliance has already been explained.[36]

Four instructions have in fact been made by the Secretary of State. All four apply to Production Licences, but only the first three apply to Exploration Licences.

(a) The Petroleum Code of Safe Practice. The first instruction is perhaps the most important. It is short enough to be reproduced in full and reads as follows:

> . . .until further notice all operations in or about the licensed area to which the licence relates shall be carried out in accordance with such provisions of Part VIII of the Institute of Petroleum Model Code of Safe Practice in the Petroleum Industry issued in October 1964 as relates to the safety, health and welfare of persons employed (in or about the exploration area).[37]

The Institute of Petroleum Code[38] is an industrial code of practice, but lacks either sanctions or real authority. It deals with 'the recommendations regarding safe practice only' and does not include anything 'issued as regulations or instructions by the appropriate national authorities'.[39] It is exactly this lack of authority which was criticised by the 'Sea Gem' Report, and led to the recommendation that 'a code of similar authority (to the US statutory regulations) supported by credible sanctions ought to be made applicable'.[40] Though very detailed in account, it is not framed as a set of instructions but rather as a lengthy advisory document. It should also be noted that many of the contractors and operators have their own safety codes or manuals of some sort which might be followed with the Code, and in many cases in preference to it. Again, there is a problem of advancing technology in compiling a detailed code of this nature which relates to the way in which specific jobs should be done. In view of the Code's importance, however, some of its contents should be noted.

The Code advises that 'all personnel employed on offshore installations should be able to swim and should have some experience of watermanship'.[41] The 'Sea Gem' Report noted that 'this may be good advice in general but it is loose in expression since no standard of ability to swim is laid down and the phrase "some experience of watermanship" is exceedingly vague'.[42] The Report also noted that this was not a requirement made of seamen on British ships, and that many of the crew of the 'Sea Gem' were in fact unable to swim. In view of the cold it is unlikely that an ability to swim would be more than limited use in the event of a collapse – a swimmer would not survive many minutes even with a lifejacket.[43] Other recommendations include use of lifejackets at all times when a rig is being raised or lowered,[44] the use of safety harness on rigs for men working aloft, working in pairs, and regular testing of the emergency evacuation procedures. The Code advises the use of warning systems, constant radio contact, gas detection alarms, the carrying of first aid kits in abundance, the admission to the rig only of authorised persons all of whom should wear safety helmets, warning notices of all restrictions (particularly those relating to naked lights) to be posted prominently, and the use of a system of permits for items of equipment, specified jobs, and specified workers. The Code is in effect a most comprehensive document. Its principal defects are that it is not statutory and that it requires adaptation for individual rigs. As far as particular application is concerned, the Code itself advises that all rigs should have their own procedures for emergencies and for particular jobs bearing the general principles in mind. The only problem is that this too is subject to the same criticism. There are no sanctions, and no enforcement procedures.

In fact the industry does make an effort to better its safety standards. Accidents are expensive, especially if they cause delay. All rigs must now have a person qualified as a nurse or in first aid on board who can discuss a particular injury or illness with a doctor on shore if it is at all beyond his ability, and all the firms have emergency arrangements with local doctors, and often with the local hospital in the event of a man having to be flown in by helicopter. Some of the newer, larger rigs, such as the 'Staflo', even have a small hospital on board, plus a comprehensive supply of drugs. On board Conoco operated rigs the crew have daily five minute safety meetings to discuss dangers and hazards that they have come across during the day, and are likely to come across in the future. Safety films are shown on board to the men, and the meetings have the effect of keeping safety in the front of the workers' minds. On their rigs too every visitor is handed a safety direction sheet which includes the emergency procedures. The rig also holds general safety meetings to discuss anything important that occurs. Abandon platform drill is held once a week, as are 'blow-out prevention' drills.[45] All the companies have safety superintendents or officers who are shore-based and visit the rigs from time to time. They carry out the most detailed inspections when they come aboard. All rigs carry lifesaving equipment, and many have Brucker escape capsules.

According to management the code of safety adhered to on board the rigs is rigidly enforced. None of the companies can afford to have accidents of the magnitude of 'Sea Gem', and the Report made it quite clear that better safety procedures than those in operation on that rig could and should be observed. The companies have taken this very seriously. Some of the companies also run safety 'incentive' schemes that involve payment to the crews if they go for certain periods of time without serious ('disablement') injuries. On the other hand, many companies have clauses in the contract enjoining the workers to observe the safety procedures. At the same time an obedience clause would render them liable to dismissal in the event of their not complying with lawful instructions. All the companies will dismiss summarily ('fly off') workers who transgress the rules. This applies particularly to smoking outside the permitted areas on board; and several instances have occurred. Statutory enforcement of the Code will no doubt go even further towards the provision of a safer system of work, but it is doubtful whether all accidents, particularly in such a hazardous industry, will be eliminated, especially as many are accounted for not by equipment failure but by simple human error in hazardous situations.

(b) Medical attention. Instruction 2 orders the licensee:

(1) to maintain contact with the Port Medical Officer or Medical Officer of Health for the British port or other base from which operations in the licensed area are being conducted;
(2) to seek advice from such Officer on. . .

(ii) the arrangements that are necessary to ensure that suitable medical attention can be provided, on an installation or on the mainland, for persons employed on operations in the licensed area.

As already indicated, the necessary facilities for treatment are available to every rig in the North Sea. Nearly all the rigs have a helicopter service to the mainland, and one of the Bristow's helicopters is available at all times in the event of emergency.

(c) Reporting of casualties. For the purposes of information the third instruction requires the licensee to inform the Secretary of State of any accident 'resulting in death or a broken limb or requiring transport of the injured from offshore to onshore for medical aid'. In addition the licensee must give details of the time and place of accident, the injured person's name, details of the accident, its causes, and the estimated period of the injured person's incapacity. The instruction also requires the reporting of any happening that could have resulted in serious injury in certain circumstances, but resulted in a lesser injury than the kind described above.

(d) Other safety instructions. Instruction 4, which applies only to Production licences, takes into account various recommendations of the 'Sea Gem' Report. It firstly requires that when any installation is anchored or affixed in connection with searching or boring for petroleum there shall always be someone on board who is the designated 'person in charge' who shall be in sole command. This is no longer relevant in view of section 4 of the Mineral Workings (Offshore Installations) Act. But the instruction also requires that the person in charge arrange for the continuous observation of any attendant vessel that is in danger of coming into contact with the installation, and for any contact to be reported to him, and for the inspection of the installation if the contact might have affected its safety. An attendant vessel had hit the 'Sea Gem' on no less than three occasions, once being sufficiently serious for the master of the vessel to make an entry in his log, but the evidence was that no one was really aware of these contacts. The Tribunal thought this was 'most unsatisfactory'.[46]

The other parts of the instruction are equally important. In particular an attendant standby vessel is required at all times at every installation whenever it is manned, unless the master of the vessel thinks that the conditions are so serious that the safety of his vessel may be jeopardised, or unless the installation is provided with effective escape capsules or other devices approved by the Secretary of State. The instruction also requires that the installation be adequately earthed, that the derrick floor have a non-slip surface, that derrick ladders should be caged, that there shall be an effective method of public address and an emergency warning system. They insist on the provision of a copy of instructions in the case of emergency to all members of the crew, and the prominent display of instructions around the installation; restrictions on smoking and naked lights; and the keeping of records of all safety matters. The last instruction also stems from a defect in information criticised by the 'Sea Gem' Report. It should be noted that provision is made for the regulation of all the matters that have formed the subject of these instructions under the Mineral Workings (Offshore Installations) Act.[47]

B. The Mineral Workings (Offshore Installations) Act 1971 and Safety Regulations

The Mineral Workings (Offshore Installations) Act, which stems from the recommendations of the 'Sea Gem' Report, is primarily deisgned to put the instructions under the licences on a regulatory footing and lay down standards for the construction of installations much like those that are made for ships. Introducing the Bill in the Commons Second. Reading Committee the then Under-Secretary for Trade and Industry, Mr Nicholas Ridley, said:

> The Bill is necessary because existing powers to control safety on drilling rigs are inadequate. They derive from the Petroleum Production Regulations and they are inadequate because they are contractual rather than mandatory. They depend upon the granting of licences to those who wish to operate rigs, but, once the licence is granted, there is weak control over the methods of operation which depends on the good will of the company operating the rig, enforcement being very difficult for the inspectors of my department who lack statutory powers. This is not to say that those who operate drilling rigs have not complied with and co-operated in all respects to increase safety, but I feel sure that the Committee will think it better to put this matter upon a statutory basis so that everyone will know what are his duties and responsibilities.[48]

As already pointed out, many of the other provisions of the Act, such as those relating to registration, discipline, and the command structure also have safety as their rationale. But the Act is concerned specifically with safety in four different ways.

1. Registration

Firstly, the Act provides in section 2 for a system of registration of installations in the British sector of the continental shelf. Whereas registration of ships is important for the purposes of ascribing nationality and applying the law of the country of registration on board, registration of an offshore installation, although based on ownership, does not have this rationale and would seem to be carried out primarily for the sake of information useful in the event of an emergency. The regulations made under this section[49] require the owner[50] of an installation to inform the Secretary of State of the type of installation, its position, what it will be doing and how long it expects to stay there, and to inform him again if it is in any way altered or moved.[51] The installation must display its registered name or other designation in such a manner as to make it readily identifiable when approached from either the sea or the air,[52] and a copy of the registration certificate must be kept posted on board where it can be conveniently read.[53] The concessionaire and the owner of the installation may be fined £200, and the installation manager £100, for a breach of the obligation to register the installation, and the owner and installation manager may also both be fined for failure to display markings or the certificate of registration. The owner may further be fined for failure to inform the Secretary of State of any changes in the particulars of registration.[54] The net result of these measures is that the Department is aware of the exact location and description of all installations in the British sector of the continental shelf. Such information might prove invaluable in the event of an accident or emergency.

2. Construction Regulations

Secondly, the Act makes it possible to impose strict standards of design and construction on installations in an attempt to prevent a repeat of the 'Sea Gem' disaster.[55] Section 3 of the Act permits regulations to be made relating to the certification of installations in respect of safety matters by competent persons,[56] surveying and inspection, and also to detail the prohibition or restrictions that may be imposed.[57] The duty to comply falls on the owner of the installation, the manager, and the concessionaire, and each may be guilty of an offence punishable summarily by a fine of up to £400, and on indictment by a term of imprisonment of up to two years, or a fine, or both.[58]

The regulations,[59] published in 1974 with an accompanying guide from the Department of Energy,[60] make it an offence from 31 August 1975 to establish a fixed installation in the waters covered by the Act, or to bring a mobile installation into those waters with a view to it being stationed there, or to maintain either kind of installation in those waters, unless it has been granted a valid Certificate of Fitness.[61] In addition, no mobile installation is to be moved to a station in any designated area unless prior to moving the owner has obtained from a competent person a report on the environmental factors at the place to which the installation is to be moved, and the owner has reasonable grounds for belief that the installation can withstand those factors.[62]

The regulations thus proceed along two separate lines. First the installation must be surveyed and approved by a competent certification authority (e.g. Lloyd's, or Bureau Veritas[63]), and in the grant of the certificate consideration will be taken of the intended location of the installation. In any event, before moving the installation the owner must be made fully aware of a list of 'environmental factors', such as water depth, tidal range, wind speeds, currents, temperatures, wave patterns, etc.[64]

The difficulty with the principal parts of the regulation, which relate to construction and design, is the comparative novelty of the structures. The various authorities still have to build up their expertise, although the rapid growth in the number of rigs being built is making this easier. The regulations require an independent assessment of the construction and design, and a survey of the installation and its equipment.[65] Surveyors are to be accorded every facility[66] and any application for a certificate must be accompanied by plans, drawings, specifications, reports and other information necessary for the surveyor to tell whether the installation complies with the requirements to the regulations.[67] These are widely drawn, and cover the stresses of the primary structure, its buoyancy and ability to weather all the conditions for which it is intended, and indeed the design and construction of the superstructure, including its resistance to waves and heavy loads, the position, strength and equipment of the helicopter pad, the watertightness of all openings and the provision and design of escape routes and living accommodation.[68] The regulations also cover the materials to be used and the standards of construction and equipment.[69] They provide for a major survey prior to the grant of a certificate[70] and for subsequent annual surveys. No repair, renovation or alteration or replacement of the installation equipment, or of its structure if it is damaged or has deteriorated so as to impair the safety, strength, stability or seaworthiness of the installation, may be carried out without notifying the relevant authority.[71] Provision is also made for the withdrawal of a certificate of fitness for a variety of reasons[72] and for exemptions.[73] By September 1975 certificates had been granted to sixty-five out of the seventy-four installations in UK waters. Only twenty-four of them received full certificates, the other forty-one were given with the proviso that new surveys were undertaken and any necessary improvements were made as a result. Nine installations were temporarily exempted. In addition, owners of twelve gas platforms in the southern part of the North Sea were told that they must evacuate all their crews when severe storms are expected as their platforms do not conform to the latest standards.

There is, of course, a large amount of flexibility in the criteria. There are a great variety of designs in construction, and it is implicit in an industry as new as the offshore industry is to the North Sea that a certain amount of experimentation will take place to tackle the adverse conditions. This is particularly so in the case of production platforms in northern waters where some form of structure has to be built to stand in 650 feet of water and to be capable of remaining there no matter what the weather. The Department is, of course, well aware of the rapidly changing nature of the technology, and for the need for the regulations and standards to be frequently adapted to take account of improvements, and the Offshore Installations Technical Advisory Committee, a body comprised mostly of technical experts, was set up to advise the Secretary of State on these matters. In February 1975 a further special committee, the Advisory Committee on Fixed Offshore Installations, was set up under the chairmanship of Sir Kirby Laing to look at the special problems of concrete structures. Nevertheless there have been warnings that on a simple statistical basis the construction materials of installations means that, like bridges, major collapses should always be expected. In fact, with more and more rigs being built this side of the Atlantic, rather than being towed across from the Gulf of Mexico and adapted for the North Sea, inspection and control over design should be easier than it was in the past, and as time goes on standards will no doubt become even higher.

3. Discipline

Thirdly, the legislation adopts the attitude that discipline is an integral part of safety on board. This is a subject which will be fully discussed in a following chapter.

4. Safety Regulations

Lastly, by section 6 the Secretary of State may make regulations for 'the safety, health and welfare of persons on offshore installations and generally for the safety of such installations and the prevention of accidents on or near them'.[74] These regulations are potentially of very wide application. They may govern people who are on the installation even if it is not part of their job to be there,[75] apply to transport of persons and things to or from the installation,[76] vessels, aircraft, or hovercraft,[77] and to any sort of work whether actually on the installation or underwater, or indeed under the sea bed.[78] A special inspectorate is provided for by the Act to supervise the execution of all the regulations made under the Act, and the Schedule, which, although expressed in general terms, does give an idea of the intentions of the legislation. The Schedule specifies measures to ensure the safety of the installation, including when it is being dismantled or assembled, and provisions as to: '. . .the manner in which or occasions on which any operation of work is to be or may be carried out, or as to the safety or suitability of any place where it is carried out'.[79] The regulations may also make provision as to the equipment, facilities, or materials which are to be or may be used,[80] or to any other matter.[81]

The importance of these regulations is beyond doubt. But even though the Act has been on the statute book for five years and was in preparation for some considerable time prior to that, the first general safety regulations have only just been published.[82] Altogether there are now several different sets.

(a) The Inspectorate. Firstly, although the central safety regulations have just been published, those covering the inspectorate appointed under section 6 (4) of the Act are already in force.[83] These give the twelve members of the Petroleum Inspectorate the power to board an installation[83a] and obtain access to all parts of it, to inspect it and indeed to require the owner or manager or any person on board or near to an installation to do or refrain from doing any act which appears to the inspector to be necessary or expedient for the purpose of averting a casualty or minimising the consequences of one.[84] The inspector has power over papers etc. and ancillary powers to ask the appropriate people for help in carrying out his functions.[85] Penalties are provided for non-compliance.[86] For example, he can demand conveyance out to the installation,[87] that equipment be tested,[88] and even demand assistance in inspecting the sea bed.[89]

The principal weakness of the inspectorate stems from the nature of the industry. As is the case with most safety inspectorates there is a problem of understaffing, but the fact that the installations can only be approached by helicopter or by boat means that to all intents and purposes it is impossible for the inspector to arrive without warning. The inspector has to book a seat on the helicopter in advance, with rig managers having something like a week's prior notice of his visit even if not directly informed. Although union representatives have asked the Department of Energy to make helicopters available, nothing has been done about this request so far.

(b) Logbooks. Secondly, the Offshore Installations (Logbooks and Registration of Death) Regulations 1972,[90] require that a logbook be kept at all times by the manager on any installation which is drilling or in the course of construction,[91] in waters to which the 1971 Act applies. Most of the entries will not relate to safety, but to the movement of personnel to and from the installation,[92] although the logbook, which must be kept in English[93] and delivered up to the Secretary of State after six months,[94] will also be used to record: '. . .every occurrence affecting or likely to affect the safety of the installation or the safety, health or welfare of persons on or working from the installation or involving the installation and endangering persons in its neighbourhood'.[95]

Apart from manning changes, visits by helicopters and any movement of the installation itself, the manager has to mention[96] adverse weather conditions, collisions, structural changes, major repairs, surveys and any other occurrence relevant to the safety, seaworthiness or stability of the installation.[97] He must also enter all safety drills, all accidents and injuries to persons, and any disease or death,[98] as well as all factors relating to emergencies, and the action taken in anticipation or as a consequence.[99] He must also mention the visit of an inspector, and if the inspector serves him a notice, or requires any action to be taken, what was done about it.[100] These requirements will ensure that the Secretary of State has full information as to what has happened on the installation in each six month period.

One criticism may be made of the procedure adopted, bearing in mind that the desire for information relating to events happening on installations is understandable, and, indeed, laudable. The manager is required to enter in the log a great number of everyday happenings within twenty-four hours of their occurrence, or as soon as possible thereafter,[101] and although the actual business of entering up the log will be long and tedious in many cases, the case for the provision of some of the information, such as manning turnover, is overriding. Nevertheless, in the event of a disaster there is no guarantee that any of this information will survive. In the 'Sea Gem' disaster this is exactly what happened. The Tribunal admitted that whatever records were kept on the installation were lost except for the 'Weekly Boring Log',[102] the log which contained all the vital commercial information as to what stage the drilling had got to, and the results of the bore. Safety records were entirely missing. It seems surprising that when the installations are in daily contact with the shore by radio-telephone, some sort of provision could not be devised for the keeping of these records on shore in a comparatively safer place.

(c) Public Inquiries and Employers' Liability Insurance. Regulations have also been made making provision for the notification of accidents and public inquiries into any accident involving loss of or danger to life which occurs on an installation.[103] These are discussed in the following chapter,[104] as are those which have been made extending the application of the Employers' Liability (Compulsory Insurance) Act 1969 to operations on the continental shelf.[105]

(d) Operational Safety Regulations. These were published in June 1976 to take effect in November. As well as introducing various manning requirements they cover the marking of hazardous areas and the use and storage of dangerous substances. They require a standardised examination, maintenance and testing routine for the installation and its equipment, and for the independent examination of all lifting gear as well as its marking. A number of general standards for equipment are also laid down, and provision is also made for the fencing of dangerous machinery. They cover drilling and production activities, provide that only essential persons should be on board the rig when moving, stipulate that certain safety equipment such as helmets and special clothing must be provided and worn, and lay down stipulations relating to scaffolding, safety harness, life-jackets, signalling equipment, helicopter operations and equipment, and institute a system of work permits and written instructions for welding and other activities where there is a potential source of ignition or risk of explosion. Apart from the imposition of a number of rather general duties on managers, concession owners and employees alike, the regulations also specify that the owner of the installation must give the manager written instructions relating to safe practices covering drilling and production, electrical and

mechanical equipment, personal procedures and general safety on board. Many gaps still remain and the standards, and specificity of the safety procedures are patchy.

(e) Diving operations. Lastly, regulations have now been published covering diving operations and activities carried out from offshore installations or relating to them.[106] They make provision for the appointment of diving supervisors to control and supervise all diving operations,[107] and impose duties upon the managers of installations and masters of vessels to provide him with information, assistance, proper facilities (including communication) and warning signals and lights to inform others of the operations.[108] The regulations prohibit diving from an installation not at anchor,[109] make provision for lifelines,[110] and prohibit dives below 50 metres without a submersible compression chamber (except in emergency)[111] and below 125 metres without in all cases informing the Secretary of State giving particulars for his approval not less than twenty-one days in advance.[112] Similar notification is needed if saturation techniques are used.[113] These duties are all imposed upon the employer and the diving supervisor. Divers themselves are also prohibited from remaining underwater for more than three hours in any twenty-four unless using saturation techniques.[114] Nor should he dive at all if he is unfit to do so.

The regulations also make provision for diving rules and tables which have to be submitted to the Secretary of State.[115] Provision is also made for a diving operations logbook[116] and a diver's logbook[117] in which all details of the dive are recorded. The diving supervisor is under a duty to make sure all operations are conducted in accordance with the diving rules,[118] that suitable standard diving equipment is used, and that standby divers are available at all times.[119] At least one must be ready to go into the water immediately,[120] and special provision is made where two divers are working together, or from a submersible compression chamber.[121] The regulations also cover the diving plant and equipment itself,[122] its testing[123] and daily examination,[124] and breathing mixtures,[125] and impose duties relating to means of communication and the provision of lights[126] and emergency services.[127] Although they also attempt to lay down certain standards for the qualifications of divers, they only stipulate that each diver has to have had 'previous experience', 'adequate theoretical and practical instruction', etc.[128] These are obviously wide criteria which demonstrate clearly that such standards are in their infancy, although the regulations themselves are forcing improvements in diving practices.

Provision is also made for medical certificates and examinations.[129]

C. The Merchant Shipping Act 1974

One of the most sensational diving accidents of the last few years (which fortunately did not result in injury) was that of the 'Pisces III', a submersible craft which was trapped on the ocean floor for three days with two men aboard. Regulations made under the Mineral Workings (Offshore Installations) Act do not cover this kind of operation, and the Merchant Shipping Act 1974 now makes provision to fill this gap. By section 17 (1):

> The Secretary of State may make regulations –
>
> (a) for the safety of submersible and supporting apparatus;
> (b) for the prevention of accidents in or near submersible or supporting apparatus;
> (c) for the safety, health and welfare of persons on or in submersible or supporting apparatus;
> (d) for prohibiting or otherwise restricting the operation of any submersible apparatus except in accordance with the conditions of a licence granted under the regulations; and
> (e) for the registration of submersible apparatus.

Wide powers are taken in relation to registration and offences,[130] but the regulations only apply to operations within territorial waters, or to apparatus launched or operated from, or comprising, a ship registered in the United Kingdom (or a British ship not so registered).[131]

The regulations themselves[132] broadly speaking duplicate those relating to operations carried out from, or in relation to, offshore installations, with suitable modifications to take account of the different type of equipment. They impose similar restrictions on diving without notification,[133] on the periods of time to be spent underwater,[134] and make similar requirements of masters of craft and diving supervisors in relation to equipment and operations.[135] Similar provision is also made for diving manuals, operations logbooks, divers' logbooks, the divers' qualifications and certificates of fitness, as well as equipment, breathing mixtures and emergency services. In addition, the regulations impose a duty to report on casualties, very much like those laid down for offshore installations.[136]

D. The Petroleum and Submarine Pipe-lines Act 1975

The Petroleum and Submarine Pipe-lines Act also makes provision for regulations to be made in relation to the safety, health and welfare of workers engaged on pipe-line works.[137] These regulations may relate to the operation of any vessel, vehicle, structure, plant, equipment or other thing for the purposes of any pipe-line works, and these are defined sufficiently widely to cover any kind of operation to do with pipes or pipe-lines.[138] Only regulations on diving have so far been made, and it should be noticed that the Secretary of State is under a duty to consult with those persons he considers representative of the persons who will be affected by them before they are.[139]

Under the Act the regulations may exclude cases or apply only to specific cases, may specify conditions which must be met before the regulations do not apply,[140] and may create offences punishable by a fine of up to £400 and by a term of imprisonment not exceeding two years.[141] The Secretary of State is empowered to appoint inspectors,[142] and regulations may be made setting out the powers and duties of the inspectorate which may include power to inspect pipe-lines, to enter upon premises, vessels or installations used or intended to be used in connection with a pipe-line or connected activities. They may also be given the power to stop operations, to require conveyance, and the regulations may also set out duties of persons required to keep records and furnish information.[143] Provision is also made for the holding of public inquiries into accidents.[144]

Although the express provisions of the Act are not as extensive as those under the Mineral Workings (Offshore Installations) Act the general nature of the power to make regulations means that the subject matter of the regulations may in fact parallel closely those made for offshore installations,as do those relating to diving.

E. The Health and Safety at Work etc. Act 1974

Although progress is thus being made on the regulation of offshore activities for safety purposes, many gaps remain. In this context it is worth pausing to reflect on the potential application of the Health and Safety at Work etc. Act 1974 to activities outside territorial waters.[145]

Although the relevant Orders have not been made as yet, the Act could apply to offshore installations. Equally, there does not seem to be any indication that this will be done in the near future. Thus in the list of relevant statutory provisions the Act is designed to effect, no mention is made of the Mineral Workings (Offshore Installations) Act 1971.[146] Nevertheless, by section 84 (3):

> Her Majesty may by Order in Council provide that the provisions of Parts I and II and this Part shall, to such an extent and for such purposes as may be specified in the Order, apply (with or without modification) to or in relation to persons, premises, work articles, substances and other matters (of whatever kind) outside Great Britain as those provisions apply with Great Britain or within a part of Great Britain so specified.

'Premises' are defined in the Act to include 'any. . .offshore installation, and any other installation (whether floating or resting on the sea bed or subsoil thereof. . .)'.[147] Such an Order in Council could apply to foreign companies and foreigners, and extend liability under section 47 (2) to acts or omissions occurring outside Great Britain.[148] If such an Order were made, then regulations made under the Act could repeal or modify any provision not contained in the list of relevant statutory provisions,[149] although it is provided in section 15 that:

> If an Order is made under s.84 (3) providing that this section shall apply to persons, premises or work outside Great Britain then notwithstanding the Order, health and safety regulations shall not apply to or in relation to. . .offshore installations outside Great Britain or persons at work outside Great Britain in connection with submarine cables or submarine pipe-lines except so far as the regulations expressly so provide.[150]

Clearly, no problems would be encountered in absorbing the offshore safety regulations under the new health and safety regulations. But no plans are being made to do so at present, and the Department of Energy's inspectorate is likely to remain separate for the time being. Indeed it is likely to remain so until the system has been completed by regulations and the process of inspection and control has become accepted. There are clear advantages at least to the companies in continuing to deal for the time being with members of the government department whose task it is to supervise the rapid development of the resources of the North Sea.

III. HEALTH REGULATIONS

Although the effects of the working conditions on the health of oil-rig workers have long given rise to concern, regulations relating to health have only just been published by the Department of Energy. But the instructions issued by the Secretary of State under the regulations made under the Petroleum (Production) Act 1934, and the Continental Shelf Act 1964, cover this area. Thus instruction 2 orders the licensee:

(1) to maintain contact with the Port Medical Officer or Medical Officer of Health for the British port or other base from which operations in the licensed area are being conducted;

(2) to seek advice from such officer on

  (i) the requirements, particularly those relating to the action to be taken in cases of infectious diseases or food poisoning, of the Public Health Act 1936, and the Food and Drugs Act 1955, or of the Public Health (Scotland) Acts 1897 and 1945 and the Food and Drugs (Scotland) Act 1956. . .

The same criticisms as were made earlier in relation to the system of instructions apply here as well. Nevertheless, the inspectors do check the rigs, and some of the companies have their own doctors who go out to the rigs once a month to check for cleanliness and to examine the hygiene. The doctor or first aid man on all the rigs can administer medicines, and if necessary take instructions from a shore-based doctor in just the same way as if the man were injured. Apart from the requirements relating to medical staff, the new safety, health and welfare regulations also require that there should be a 'sick bay' on each installation, and detail the medical stores and medicines which must be carried on board. They also require clearly marked sources of drinking water, and lay down a vague general standard for provisions. 'Welfare' in the regulations is supposedly covered by the imposition of a minimum age of 18 for employees, and the requirement that a strict record be kept of all hours worked.

IV. ACCOMMODATION AND RECREATION FACILITIES

Most companies provide similar facilities for the men out on the rigs. The living quarters are nearly all designed the same way – as a cabin structure, usually self-contained and capable of being dismantled in the case of some jack-ups, situated at one end of the installation away from the drilling, but generally below the helicopter deck. The physical structure invariably contains a mess room, a recreation room of some kind, the kitchens, showers and the bedrooms as well as the administration offices. Men usually sleep 4-6 to a cabin in bunks, though in some cases this figure is higher, sometimes as many as twelve or fourteen to a room.[151] On the Shell rig 'Staflo', a very large semi-submersible, the men graduate to smaller rooms as they go up the hierarchy of command. Thus most sleep six to a cabin, although the toolpusher and senior staff have cabins to themselves.

In fact these remarks obscure what has become a source of vociferous complaint amongst offshore oil-rig workers. This is the operation of the so-called 'hot-bed' principle on some of the older rigs. Some offshore rigs provide accommodation for only one crew – the shift off duty. When the shifts change, the other crew use the same accommodation. Two men therefore sleep in the same bunk, each for half the day. It must be mentioned too that not all the bunks are very large – some are as small as two feet across. In this case comfort has been sacrificed to the demands of space. Even if this is not the case, however, it must be remembered that the rig is in constant operation. Noise is a problem that many cannot come to terms with.

By contrast, or by way of compensation, most companies demand from their independent caterers that food be of first-class hotel standard. The investment, for the food is free, is regarded as a good one. There is little else to look forward to on the rig, and it is undoubtedly physically necessary. It also reduces the area of dissatisfaction and conflict, a policy of great importance in the administration of total institutions. One must add that, apart from the BP rigs, most installations are 'dry'. Alcohol is thought by many to be dangerous in conditions of isolation,[152] and the risk of hoarding, or of passing on to others, is considered to be too great.

All the companies show films on their rigs. Shell show two top-line films per week, though not on all production rigs. Most rigs have games rooms with billiards, cards, magazines, darts, TV and so forth and in some cases a small 'library' (a bookshelf of paperbacks). In general the men are well provided for, considering that the twelve hour shift leaves little time for recreation, and the men expect to work flat out in their shift period. As newer rigs get built oil men can expect higher standards. The North Sea is proving tough on the older installations brought over from the USA, and the ones that have been built for the North Sea, such as the 'Staflo' and 'Sea Quest' are all of a much higher standard. But the overall effect of these improvements should be matched against the rate of labour turnover, in much the same way as the test of safety regulations is the number of accidents that occur nevertheless. Just as danger cannot be eliminated from North Sea operations, so they cannot be expect to be conducted without discomfort, but steps are being made and should be pursued, to remove as much of both as possible.

6 THE LEGAL CONSEQUENCES OF OFFSHORE ACCIDENTS

'Publicity is justly commended as a remedy for social and industrial diseases. Sunlight is said to be the best of disinfectants; electric light the most efficient policemen.' Louis D. Brandeis, *Other People's Money,* Frederick A. Stokes Co., New York, 1914, p.92.)

Plainly, despite preventative measures and protective legislation, accidents often happen on offshore installations. An employee who is hurt may be treated on the rig, or flown off for medical attention ashore if the matter is more serious. He may return to work relatively soon, but in some cases his injuries, if they did not result in death, may mean that he suffers pain, loss of employment and opportunity. The accident may well not have been his fault. In all cases there are some legal consequences whether they be as insignificant as the duty on the installation manager to record the incident or as serious as the employee or his relatives mounting a claim against his employers for damages.

As already outlined in Chapter 2, section 3 (2) of the Continental Shelf Act 1964 (extended by section 8 of the Mineral Workings (Offshore Installations) Act 1971), makes it possible to bring actions in tort based on 'acts or omissions taking place in a designated area. . .in connection with the exploration of the sea bed or subsoil or the exploitation of their natural resources. . .'. The regulations give the various courts power to determine such questions as if the acts or omissions had taken place within their jurisdiction on land. This has several results, and the options open to an injured employee are, broadly speaking, as wide as those open to a worker injured in the petroleum industry on shore.

In addition, there are certain duties relating to notification imposed on the employer of a person who is killed or injured on an offshore installation. There are also special provisions relating to inquiries and investigations as well as those pertaining to social security. All of these consequences merit attention.

I. COMMON LAW

The employer of workers on an offshore installation, be he an oil company or an offshore exploration company, a drilling concern, a catering firm under contract to supply the food and 'housekeep' on board, or an operating company in charge of the rig, is under the normal obligation of the common law which 'is threefold, "the provision of a competent staff of men, adequate material, and a proper system and effective supervision" '.[1] In other words he is subject to the ordinary common law duty to take reasonable care of his employees. It is not appropriate to consider here fully a subject adequately covered elsewhere.[2] But some of the supplementary rules to this duty might usefully be mentioned in relation to the specific employment in question, for in actions for negligence the standard of care might well be crucial.

It must be borne in mind at the outset, however, that although claims based upon the breach of the duty of care are normally considered to fall within the province of the law of tort, they may equally arise as claims for breach of an implied term of the contract.[3] This may well be crucial for an offshore oil-rig worker when questions of jurisdiction arise; for example, when the rig is just over a median line, or even out of a designated area. In *Matthews* v. *Kuwait Bechtel Corporation,*[4] an employee working in Kuwait under a contract expressly stated to be governed by English law, was injured in the course of his employment. The court held he could claim damages in tort or for breach of contract, and in respect of the latter claim, service out of the jurisdiction was allowed. Bearing this in mind, it is proposed nevertheless to discuss the common law duty of care only in relation to tort, remembering that even in cases brought under contract, negligence must still be proved. There are indeed several facets of the law of negligence which are of particular relevance here.

Firstly, if one employment is more dangerous than another, a greater care must be taken.[5] Though said in the course of an action by a workman injured in his one good eye, the general statement would enlarge the duty of care to be exercised on an offshore oil-rig. The dangers are obviously much greater. But there are qualifications to this approach. In a hazardous occupation where risks are usually present it is not necessarily negligent for the employer to misjudge the risks. The usual example is of the deep-sea fishing industry, where a skipper kept men at work cleaning the fish deck in rough seas, and one was injured by an exceptionally heavy wave.[6] In cases of rigs being jacked up in heavy weather to avoid being buffeted by enormous waves the situation is arguably analogous.

In dangerous occupation there is also the threat of *valenti non fit injuria,* the defence of assumption of risk.[7] In *Bowater* v.*Rowley Regis*

Corporation[8] Goddard L.J. said: '. . .it can hardly ever be applicable where the act to which the servant is said to be "volens" arises out of his ordinary duty, unless the work for which he is engaged is one in which danger is necessarily involved'.[9] In the later case of *I.C.I.* v. *Shatwell*[10] the House of Lords expressly approved the general principle that assumption of risk will not ordinarily defeat a claim by an employee against his employer, though it did on the facts of that particular case. The question of whether it would in cases concerning offshore installations is clearly a problem of tort that goes beyond this study, to all forms of dangerous occupations, and one which is related to the nature of the negligent act and the particular circumstances that surround it. But it is certainly the case that oil-rig workers are appreciative of the risks involved.

Secondly, it must be remembered that the duty of care laid down in *Wilsons'* case is a personal one, and cannot be discharged by entrusting the provisions to competent delegates.[11] But the liability is one for negligence, and not one of absolute liability, and the question is always one of whether the employer acted or failed to act in such a way as to constitute a lack of reasonable care.[12] In view of the complicated working arrangements between different companies that exists on many offshore rigs it is worth considering the position of independent contractors. The proposition is simply made. If the duty of care is a personal one on the employer, it is not discharged by the appointment of an independent contractor to carry out the work, any more than it is possible to discharge the duty by asking a competent employee to perform the task. Thus if an independent contractor is engaged to perform some work essential to the operations, and there are many conceivable examples of such situations offshore, the employer may well be delegating part of his duty. But he then remains liable for any lapse of reasonable care on the part of the persons doing the work on his behalf.[13]

Perhaps more important, therefore, in this context, is the vicarious liability of an employer for negligent acts committed by one of his employees in the course of his employment which causes damage to another worker or person on the rig. Again, this is adequately covered elsewhere,[14] but, accepting the principle, the exact meaning of 'in the course of employment' is worthy of special consideration, especially as the duty of the employer to provide the safe system of work is subject to the same limitation.

In *Davidson* v. *Handley Page Ltd.*[15] Lord Greene, M.R. said: 'The obligation of the employer extends to cover all such acts as are normally and reasonably incidental to a man's day's work.'[16] In that case the plaintiff had gone to wash out a tea cup when she slipped on an oily duck-board and injured herself. The difficulty with those who are engaged in 'total' occupations is that the distinction between work and 'non-work' is one only capable of complete definition at the end of the worker's period away from home, and in the case of the oil-rig worker, out on the rig. One is faced with the problem, still undecided in law, of the shift worker who stays at his place of work, under the contract though not working, after his shift is over. In the case of the oil-rig worker, he is, like others in total occupations such as merchant seamen, physically incapable of leaving. The staying is in fact part of the job. If he is injured whilst off duty in such circumstances the question remains: was the injury 'in the course of his employment?'. The answer appears to be unclear, but authority seems to favour the view that in the case of employed persons 'whose service is continuous. . any accident occurring to them during the period of their engagement must arise in the course of their employment'.[17]

In fact, the answer may well be found, not in this notion of continuous employment at all, but in arguments advanced as to the casual connection between the accident and the employment.[18] The clearest statement of this approach is to be found in *St. Helen's Colliery* v. *Hewitson*[19] where Lord Atkinson said that a worker is acting in the course of his employment when he is: '. . .doing something he was engaged to do. . .or. . .when he is doing something in discharge of a duty to his employer, directly or indirectly, imposed upon him by his contract of service'.[20] The problem still arises with accidents which occur in situations that are natural corollaries of being on the rig for twenty-four hours a day, and thus might be said to be 'reasonably incidental' to the employment.[21] But cases like shutting someone's fingers in the firedoor leading to the recreation room might escape even this extension. Another principle decided by the courts might be of importance in considering liability: that the worker does not cease to be in the course of his employment by the fact that he is working in an area where he is forbidden to go,[22] even if forbidden by a statutory order.[23] This might have considerable impact where the regulations made under the Mineral Workings (Offshore Installations) Act 1971 concern the control of certain actions for the sake of safety.

Leaving aside the question, well beyond the scope and intention of this study, of when the worker is a 'servant' so that the vicarious liability can be ascribed to the 'master', one interesting problem remains, well illustrated on oil-rigs: the problem of the 'loaned' employee. As already indicated, some of the companies involved in the North Sea if not themselves large international concerns are at the very least served by other companies who are. The best example is the Shell group of companies who may move an employee from their Dutch company to their English company if his expertise is needed for a short period, or if that is the only way that the employee in question can pursue his designated 'career plan' moving from one job to another. The incidence of such arrangements would appear to be high enough for the situation to be described as common. In the case of an accident being caused by the employee on loan the question is which of the companies will assume responsibility for the action, and indeed if the employee is himself injured, whose duty of care, if any, was broken.

The test as to who is liable as between the 'borrowing employer' and the 'lending employer' is complicated by considerations of contract. It is therefore important to note first that a contract of service cannot be assigned unilaterally.[24] The agreement of the worker must be sought, which means either a variation of the contract or a new contract. As Lord Denning once put it:

> Much of the difficulty which surrounds this subject arises out of the nineteenth century conception that a servant of a general employer may be transferred to a temporary employer so as to become for the time being the servant of the temporary employer. That conception is a very useful device to put liability on the shoulders of the one who should properly bear it. But it does not affect the contract of service itself. No contract of service can be transferred from one employer to another without the servant's consent: and this consent is not to be raised by operation of law but only by the real consent in fact of the man, express or implied: see *Nokes* v. *Doncaster Amalgamated Collieries, Ltd.* [1940] A.C. 1104. In none of the transfer cases which have been cited to us[25] has the consent of the man been sought or obtained. The general employer has simply told him to go and do some particular work for the temporary employer and he has gone. The supposed transfer when it takes place is nothing more than a very convenient and just device to put liability on to the temporary employer. In recent years this device has been very much restricted in its operation.[26]

But if nothing is said by the employer or the worker about the contract, and the worker is put to work for someone else, one has to fall back on the common law principles. Here a distinction is to be drawn between lending a worker and making his service available,[27] though this distinction itself seems to turn on the question of control. It is the right of control over the worker which carries with it the burden of the duty of care.[28]

Though the old cases accepted that a servant cannot have two masters at the same time,[29] because only one can be responsible for the control of his work at once, the situation has become much more complicated as the test of who has control would now appear to be a multiple one. The leading case remains *Mersey Docks & Harbour Board* v. *Coggins & Griffiths (Liverpool) Ltd. and McFarlane,*[30] and in simple terms the question appears to be whether the accident occurred in the exercise of a discretion as to the manner in which the work should be performed vested in him by one or the other employing companies. The burden of proving that control over the employee has passed from his usual employer to the temporary employer is a very heavy one,[31] and is not usually discharged in the case of an employee loaned with equipment he is to use or control (this might cover a diver with a diving bell), or in the case of an expert.[32] Transfer of control is only really thought of as possible in relation to unskilled or semi-skilled workers, such as electricians[33] or labourers[34] lent on their own without equipment. As the likelihood of roustabouts being transferred in this way is small there is all the more reason to expect that the company which offered the worker his contract of employment will nearly always remain the liable company. This is especially so as the categories of offshore worker transferred between companies are almost always professional, being either mud engineers, petroleum engineers, and other 'experts'.

II. CONFLICT OF LAWS

The possibility of problems of conflicts of laws arising out of claims in tort by an offshore oil-rig worker are not so great as those arising out of contract. This is for several reasons. Firstly, section 3 (2) of the Continental Shelf Act has rendered null the possibility of arguing that for the purposes of an action the continental shelf is abroad. Secondly, in respect of torts committed in the Scottish sector of the continental shelf actions may be brought in Scotland. Thirdly, though liability for industrial injuries is primarily envisaged as springing from a duty within the province of the law of tort, a claim by an employee against his employer for failing to provide a safe system of work can, following *Matthews* v. *Kuwait Bechtel Corporation,*[35] alternatively be brought in contract. Nevertheless, the general principles should be detailed to place the problem in perspective.[36]

The English courts have adopted a somewhat idiosyncratic stance on conflict in torts.[37] The test is one that derives from the decision of Willis J. in *Phillips* v. *Eyre.*[38] Prior to the confused decision of the House of Lords in *Chaplin* v. *Boys,*[39] the substance of the rule was that a tort committed in England was entirely governed by the law of the place where the cause of action arose (being identical with the law of the place where the action was brought) and that as a general rule a tort committed abroad was actionable in England only if it was actionable under the law of England and no defence to the action was provided in the law of the country where the accident happened. The decision in *Chaplin* v. *Boys* has thrown the picture into disarray, but as a majority of the House seemed to prefer a double actionability rule, the act complained of must be a tort by English law and also actionable under the civil law of the country where the accident happened. The application of the rule is flexible, however, and indeed necessarily so to prevent 'forum shopping'. The most obvious result is that as far as an action mounted in England is concerned, the nationality of the worker, his company or his contract of employment is not as important as it is in contract.

It would not be appropriate to discuss the decision in *Chaplin* v. *Boys* in detail here. But mention must be made of the problem raised in *Sayers* v. *International Drilling Co. N.V.,*[40] although that too has been adequately treated elsewhere.[41] The problem can be stated in the following way: in an action brought in respect of a tort committed abroad, the defence might be raised that the contract between the parties excluded or modified the liability of the defendant. If the proper law of the contract and the law of the country where the act or omission complained of occurred are different, there is always the possibility that the defence may be valid according to one and invalid under the other. In such a situation which law does the court apply? If the action is brought in contract, the proper law of the contract determines the validity of the clause, subject to questions of public policy and legislation which, at least in relation to personal accidents, has been held to apply to acts occurring within the jurisdiction of the courts hearing the action even if the law governing the contract would exclude it.[42]

But if the claim is brought in tort the situation is further complicated. In *Sayer's* case Lords Justice Stamp and Salmon considered the question of the proper law of the contract as determining the validity of the exemption clause. Lord Denning M.R. examined the proper law of both the claim and the defence and asserted that only one law, the law with which the issues had their closest connection, should be applied. This he too considered to be Dutch. As the majority did not consider the tort aspect very thoroughly, and in particular ignored the fact that the claim was in tort, no consideration was given to *Chaplin* v. *Boys,* and the law of the place where the accident occurred was ruled out without consideration. Had they followed the line of reasoning of the rules of conflict of laws of tort, it is submitted that a somewhat different decision might have been reached, and one which might have clarified rather than exacerbated an already confused area of complex legal rules. A more recent Scottish case has held in a rather similar situation that where in respect of an alleged wrong committed in Scotland an action is brought in the Scottish courts and under Scots law a statute would oust a contractual defence, then Scots law should prevail over the law of the contract and the defence should not succeed.[43]

III. ACTIONS FOR BREACH OF STATUTORY DUTY

The action for breach of statutory duty is not to be confused with the action for negligence against an employer.[44] In this case the duty is one imposed on the defendant employer[45] by a statute or regulations made under a statute. Normally one would also need to ask the question: has the Act given a right of action in tort?[46] In the case of offshore oil-rig workers the question is already answered by the Mineral Workings (Offshore Installations) Act 1971. Section 11 provides:

> (1) This section has effect as respects –
>
> (a) a duty imposed on any person by any provision of this Act, or
> (b) a duty imposed on any person by any provision of regulations made under this Act which expressly applies the provisions of this section.
>
> (2) Breach of any such duty shall be actionable so far, and only so far, as it causes personal injury. . .

In relation to those engaged in pipe-line works, similar language is used to the same effect in section 30 (1) of the Petroleum and Submarine Pipe-lines Act 1975 in respect of breaches of any duty imposed by regulations made under that Act.

In the case of workers on installations, any breach of a duty set out in the 1971 Act, or set out in any regulations made under it which specifically mention section 11, causing personal injury, could be actionable by the injured person. The duties set out in the Act relate to the appointment of managers,[47] the powers of managers, the actions that the manager shall not permit where safety might be endangered, and the duty to comply with the regulations relating to certification testing, inspection and modification of installations for the purpose of obtaining a certificate of fitness. The duty of an owner of an installation to ensure that the installation manager does not permit any operations to be carried out from the installation that might endanger its seaworthiness and stability[48] is the most important in the present context, and is plainly directed at incompetence producing injuries to any of the workers on the rig. The diving regulations made under the Act also state that section 11 is to apply in relation to any of the duties which they impose on any person,[49] as do the safety regulations.

In such circumstances the owner or employer charged with a breach of statutory duty cannot plead devolvement as a defence,[50] and the same applies if the duty is neglected by an independent contractor.[51] But the injured worker in an action for breach of statutory duty has to show that the harm he suffered was within the class of risks at which the statute was directed,[52] and that he was one of the people intended to be protected. This latter provision should cause no difficulty in view of the expressed intention of section 6 that regulations are for 'the safety, health and welfare of persons on offshore installations in waters to which this Act applies'. Though not all the regulations will be made under this section, those relating to safety (and therefore the more likely to be connected with events that might lead to accidents and personal injury), certainly will be, and it is section 6 (3) which confers the general powers in this respect. The worker will still be subject to the usual rules relating to causation, which have caused difficulty in some cases.[53]

An interesting point arises with the definition given in section 11 (7) of 'personal injury'. It resembles that used in other statutes,[54] and is a humane definition capable of covering as yet unidentified conditions that might arise in a new industry such as the one in question. The subsection reads: 'In this section "personal injury" includes any disease and any impairment of a person's physical or mental condition and includes any fatal injury.' Whether regulations relating to hours and the relief of men on the installation[55] would give rise to an action based on the impairment of a worker's mental facilities caused by the isolation and fear is difficult to determine. An action on these facts in negligence would probably encounter the defence of *volenti,* but this would not be the case if it could be shown to be attributable to a breach of the regulations.

The Act also makes two further concessions as regards actions for breach of statutory duty. Firstly, by section 11 (4) it provides that none of the defences available to the employer, manager or owner in criminal proceedings by virtue of section 9 (3) or in regulations made under section 7 (2) (b)[56] shall afford a defence in any civil proceedings. Thus the use of all due diligence to enforce the provisions of sections 3, 4 and 5 of the Act is of no avail should breach of the statute cause personal injury. This remains so even if the contravention was committed without the 'consent, connivance or wilful default' of the defendant.[57] This emphasises the personal nature of the duties in question, and the fact that they are almost, in this case, duties of strict liability.

The second concession is the proviso contained in section 11 (3) that the action permitted by section 11 (2) is 'without prejudice to any action which lies apart from the provisions of this Act'. The two possible options relevant here are discussed below.

IV. OCCUPIERS' LIABILITY

As section 6 (2) of the Mineral Workings (Offshore Installations) Act specifies that the regulations made for the safety, health and welfare of persons on offshore installations may have effect as respects persons 'whether or not present in the course of their employment', the fact that there might even occasionally be present on board persons who are not employees of the owner or operator makes it necessary to consider the duty owed to such persons in case an accident should happen to one of them. In such a case the duty owed by an occupier of premises to his visitor is of great importance.

The provisions of the Occupiers' Liability Act, 1957, have effect: '. . .in place of the rules of the common law, to regulate the duty which an occupier of premises owes to his visitors in respect of dangers due to the state of the premises or to things done or omitted to be done on them'.[58]

As section 3 of the Continental Shelf Act, as extended, brings questions arising out of acts or omissions within the jurisdiction of the courts of the United Kingdom it would appear that following the arguments of Chapter 2, the Act might well apply to offshore installations and indeed to pipe-laying barges. Indeed, section 1 (3) of the Occupiers' Liability Act goes on to say:

> The rules so enacted in relation to the occupier of premises and his visitors shall also apply, in like manner and to the like extent as the principles applicable at common law to an occupier of premises and his invitees and licensees would apply, to regulate –
>
> (a) the obligations of a person occupying or having control over any fixed or movable structure, including any vessel, vehicle or aircraft,
>
> (b) the obligations of a person occupying or having control over any premises or structure in respect of damage to property, including the property of persons who are not themselves his visitors.

It is submitted that both offshore installations and pipe-laying barges fall within the definition of 'premises' or 'fixed or movable structures'.[59] Ships, of course, are obviously within the Act, and it is arguable that, as some installations are also 'vessels' the others ought to fall within what is clearly the general term covering all such places of work.

In the complicated industrial relationship of the various persons on an installation, or in control of parts of an installation, with different companies being responsible for different aspects of the work (the operating company in charge of the place it drills at, the drilling company of the actual drilling, and the catering company of the household arrangements and staff), it is vital to be able to identify the 'occupier'. In fact it is clear from the cases that there might be more than one occupier at any one time, and that exclusive occupation of the premises is not necessary.[60] The test of who is an occupier appears to be one of whether a person has some degree of control associated with and arising from his presence in, and use of, or activity in the premises. As Wrottesley J. said in *Duncan* v. *Cammell Laird & Co. Ltd.*:[61]

> It seems to me that the importance of establishing that the defendant who invited is the occupier of premises lies in the fact that with occupation goes control. And the importance of control is that it affords the opportunity to know that the plaintiff is coming on to the premises, to know the premises, and to become aware of dangers whether concealed or not, and to remedy them, or at least to warn those that are invited on to the premises.

It is therefore the *control* of the premises that decides who is liable as an occupier.[62]

If this is so it would be of great importance to know, in the case of the complicated arrangements on installations, exactly where the accident happened. In the light of the cases it might be that an accident in the kitchen would be the liability of the catering company, and an accident on the platform floor that of the owner. But this would depend on the facts of each case. Certainly non-exclusive occupation is not a stumbling block to liability.[63] But the power to exclude persons from the premises is taken as an indication of possession of control.[64] This being the case it is more likely than not to be the firm operating the rig, and providing the installation manager, who will have control, and therefore occupation.

It is necessary, too, to know at whom the duty of care owed by the occupier is directed. The physical nature of the installation makes it extremely unlikely that the owner will ever have to suffer trespassers or even 'licensees' under the old common law. Clearly the category of 'visitor' under the Act covers all those who might get on to a rig by helicopter or by boat – this is especially so in view of the powers of the master with relation to surrounding transport. Thus dignitaries and others invited out to have a look, visiting members of other installations and so forth would all fall in the category – anyone who is allowed onto the rig, whether or not invited, is covered by the Act.[65] So too are those persons who are not even welcome, but present by virtue of a right conferred by law.[66] This would cover government safety inspectors, for example.

The actual duty of care owed by the occupier to all these persons is: '. . .such care as in all the circumstances of the case is reasonable to see that the visitor will be reasonably safe in using the premises for the purposes for which he is invited or permitted by the occupier to be there'.[67] It is a question of fact in each case whether the defendant has observed the standard of care required by the subsection.[68] More important, if the visitor does not use the premises for the purposes for which he has been admitted, then he is outside the statute. But liability can be excluded or disclaimed in other circumstances as well. For example, a warning of the danger might well absolve the occupier. Section 2 (4) (a) provides: '. . .where damage is caused to a visitor by a danger of which he had been warned by the occupier, the warning is not to be treated without more as absolving the occupier from liability, unless in all the circumstances it was enough to enable the visitor to be reasonably safe.'

Though this section poses its own problems of interpretation,[69] it might enable the occupier to disclaim responsibility under the Act.

Some companies[70] hand all visitors to their rigs directions for safety, including procedure in case of alarm, and all rigs have numerous safety warning signs and danger signs in the requisite places. Whether this would be sufficient depends on the facts of each case, but the companies do at least make an effort. It must be remembered too that risks which can be shown to have been 'willingly accepted' by the visitor are excluded.[71] In other words, the doctrine of assumption of risk applies, and in the case of a visit to an offshore installation, some of the hazards of the job are bound to have been assumed, though the Act does not follow the common law principle (and the recommendation of the Law Reform Committee[72]) that no duty of care is owed when the visitor has full knowledge of the nature and extent of the danger.[73] The silence of the Act on this matter, apart from assumption of risk and warnings, would seem to result in the knowledge of the visitor being relevant to the decision as to whether due care has been taken if this awareness is itself known to the occupier.

The occupier is also free to exclude liability by agreement or otherwise,[74] and it seems that warning notices, if sufficiently brought to the visitor's attention in this respect, will serve this purpose.[75] This right of exclusion does not extend to visitors who are third parties to a contract for the use of the premises.[76] The rights of visitors of the operating company thus cannot be excluded in the contract between the operators and the contractors (owners), and the same would apply to persons who enter on the installation under a contract between the operators and a delivery company.[77]

The occupier may also escape liability in circumstances where it was reasonable for him to have employed an independent contractor to do the work, the faulty execution of which caused the damage to the visitor. Section 2 (4) (b)[78] provides:

> Where damage is caused to a visitor by a danger due to the faulty execution of any work of construction, maintenance or repair by an independent contractor employed by the occupier, the occupier is not to be treated without more as answerable for the danger if in all the circumstances he acted reasonably in entrusting the work to an independent contractor and had taken such steps (if any) as he reasonably ought in order to satisfy himself that the contractor was competent and that the work had been properly done.

The cases [79] show that it is reasonable to employ an independent contractor where the work requires a special skill or equipment that the occupier does not have. In relation to repairs to offshore installations, for example to electrical equipment, this might be of importance, and the owners may frequently need the assistance of marine construction engineers to undertake repairs. Delegation seems to be acceptable in cases where it is found to be the normal commercial practice, though this is a question in fact, as indeed is the principle question of whether the duty of care has been discharged or not.

V. THE FATAL ACCIDENTS LEGISLATION

If a worker is killed in a situation where, had he been only injured he would have had an action for damages in negligence, his dependants can mount an action under the Fatal Accidents Acts 1848 to 1959.[80] The cause of death must be due to the 'wrongful act, neglect or default' of the defendant, and the Mineral Workings (Offshore Installations) Act 1971, section 11 (2) reads into these words any breach of duty imposed on any person by any provision of the Act or any regulations made under the Act.[81] Thus death from breach of statutory duty is also actionable under the Fatal Accidents legislation, and as already indicated, the liability in such cases will be strict. It must also be remembered that any claim for loss of expectation of life that the deceased would have had survives for the benefit of his estate under the Law Reform (Miscellaneous Provisions) Act 1934.[82] The Damages (Scotland) Act 1976 provides for similar actions in Scotland.

The Fatal Accidents Act is also applied to deaths of passengers in aircraft, which might prove important to oil-rig workers whose main method of communication with the shore is by helicopter. Section 3 of the Carriage by Air Act 1961 now provides that references in the Fatal Accidents Act 1976, section 1, to a wrongful act, neglect or default shall include references to an event which gives rise to liability under Article 17 of the Warsaw Convention, as amended at the Hague in 1955. This has the effect of making the carrier liable for damage sustained in the event of the death of a passenger if the accident which caused the death took place on board the aircraft or in the course of any of the operations of embarking or disembarking.[83] Liability is not dependent on proof of fault, which goes further than any other action under the Fatal Accidents Act.[84] There is some doubt as to whether the limitation period and the measure of damages will be the same either. The Convention provides that the limitation period shall be two years,[85] and that the carrier's liability is limited to 250,000 gold francs[86] for each passenger unless a higher limit is agreed.[87] Contributory negligence also applies.[88] Of slightly wider application is the Civil

Aviation Act 1949, under which damages are recoverable without proof of negligence (unless there has been contributory negligence) where death is caused to any person on land or water by, or by any person in, or an article or person falling from, an aircraft in flight, taking off, or landing.[89]

VI. EMPLOYERS' LIABILITY INSURANCE

Under the Employers' Liability (Compulsory Insurance) Act 1969 an employer carrying on business in Great Britain is under an obligation to maintain insurance against liability for bodily injury or disease sustained by his employees and arising in and out of the course of their employment in Great Britain. Except as far as regulations otherwise provide the obligation does not extend to the coverage of injury or disease suffered or contracted outside Great Britain.[90]

Without modification by regulations therefore, the Act would not apply in respect of operations on the continental shelf with possibly unfortunate consequences particularly for employees of small sub-contracting firms who suffered serious injuries. Regulations have now been made, however, extending the application of the Act,[91] and in relation to employees who are ordinarily resident in the UK, or, if not ordinarily resident in the UK, have been present in the UK and waters to which the Mineral Workings (Offshore Installations) Act applies for a continuous period of not less than seven days in the course of employment there (called the 'relevant' employees[92]), section 1 (1) of the 1969 Act now reads:

> . . .every employer shall insure and maintain insurance under one or more approved policies with an approved insurer or insurers against liability for bodily injury or disease sustained by those of his relevant employees who are employed by him for work on or from a relevant installation, or on or from an associated structure in the course of any operation undertaken on or in connection with a relevant installation, and arising in and out of the course of their employment for that work.[93]

A 'relevant installation' for the purposes of the regulations means an offshore installation in waters to which the Mineral Workings (Offshore Installations) Act applies outside the UK other than territorial waters adjacent to Northern Ireland, not being an installation registered as a vessel which is a dredging installation or is in transit to or from a station. An 'associated structure' means a vessel, aircraft or hovercraft attendant on the installation or any floating structure used in connection with the installation.[94]

It should be noted that if on any day an employer is not insured in accordance with the Act the concession owner and the owner of the installation are both guilty of an offence punishable by a fine of up to £200.[95] Further regulations made under the 1969 Act itself effectively provide that the employer of the persons referred to is to make arrangements to maintain copies of the certificate of insurance on the installation or associated structure on or from which those persons are employed and to produce it to them on request.[96]

VII. SOCIAL SECURITY BENEFITS

From the inception of the licensing system, special provisions have been made for those persons working on the continental shelf to be covered by the social security system in relation to accidents.[97] The overhaul of the legislation now principally consolidated in the Social Security Act 1975 has made no difference in this respect, and section 132 of the Act provides that:

> (1) The Secretary of State may by regulations provide for modifying Parts I, II and III of this Act, in such manner as he thinks proper, in their application to persons in any prescribed employment (whether under a contract of service or not) in connection with continental shelf operations.
>
> (3) In particular (but without prejudice to the generality of subsection (1) above), the regulations may provide for any prescribed provision of Parts I and II of this Act to apply to such a person notwithstanding that he does not fall within the description of an employed or self-employed earner, or does not fulfil the conditions prescribed under section 1 (6) of this Act as to residence or presence in Great Britain.

Accordingly, under regulations made under this section and section 51 of the Act:

> Employment in any designated area which is employment in connection with the exploitation of the resources mentioned in section 1 (1). . .of the Continental Shelf Act 1964 or with the exploration of the sea bed and subsoil in any designated area and which, were every such area in Great Britain, would be employed earner's employment under Chapter IV of Part II of the Act of 1975'

is to be 'employed earner's employment' for the purposes of granting entitlement to benefits in respect of personal injuries and prescribed diseases arising out of and in the course of such employment.[98]

Naturally, the fact that the continental shelf is outside Great Britain has necessitated a number of modifications for which power is given in the Act. The regulations thus permit benefit to be paid notwithstanding that the accident happens or the prescribed disease is contracted while the person in question is outside Great Britain, and indeed benefit is payable if the accident happens or the disease is contracted at a time when the person is travelling from one designated area of the shelf to another or while going to or from Great Britain to or from a designated area.[99] But the application of phrases such as 'arising out of and in the course of employment' has not been interpreted in relation to oil-rig workers, at least in reported decisions, and this limitation may be important when such workers are injured on an installation when not actually working.

Although the regulations have not altered the application of the various sections of the Act relevant to some of the special situation in which accidents might occur offshore, the fact that an employee is not barred from claiming in respect of an accident which occurs when he is in breach of any statutory or other regulations[100] may be very relevant in relation to safety regulations made under the Mineral Workings (Offshore Installations) Act and the Petroleum and Submarine Pipe-lines Act. The extension of coverage of the Social Security Act to accidents happening while the employee is travelling either by boat or helicopter to or from the installation is also of enormous importance.[101] The oil-rig worker has no option but to go to work in this way, and in all cases the service is provided under arrangements made by the employer. Indeed, it is often a specific condition of employment that the employee should be willing to travel by air, thus clearly bringing him within the provision.[102]

The Act also provides coverage for accidents which happen while the employee is attempting to rescue someone, or in the course of attempting to meet an actual or supposed emergency on the installation.[103] It should be noted, too, that specific provision is made in the regulations to maintain the right to benefit of an employee injured while working on the continental shelf if he would normally have been disqualified because by being outside Great Britain he was unable to perform an act required to be done by the legislation or the regulations forthwith or within a specified time or on the happening of a certain event. In the case of oil-rig workers, performance 'as soon as is reasonably practicable' after the time or event is good enough if it would otherwise have been impossible due to the employee's absence from Great Britain.[104] The special provisions of the Act relating to reciprocity with other countries[105] are also of great importance to those working in the North Sea, not only because some may be foreign nationals, but also because they may move across median lines from time to time.

Should an oil-rig worker fall sick, as distinct from being injured, he may also be able to claim sickness benefit because in fact the regulations extend not only the provisions relating to injury, but the scope of the entire social security system to workers employed in designated areas of the continental shelf. Thus under the regulations relating to contributions prescribed employment is more or less the same as it is for the purposes of claims relating to accidents and prescribed diseases, namely:

> . . .any employment (whether under a contract of service or not) in any area which may from time to time be designated by Order in Council under section 1 (7) of the Continental Shelf Act 1964, where the employment is in connection with the exploitation of the resources mentioned in section 1 (1) of that Act (exploration and exploitation of continental shelf) or with the exploration of the sea bed and subsoil in such designated areas.[106]

In this respect it is interesting to note that the regulations proceed on the basis of a fictional assimilation of the designated areas into Great Britain. Thus Part I of the Act and so much of Part IV which relates to contributions applies in respect of a person in prescribed employment on the continental shelf, 'as though the area so designated as aforesaid were in Great Britain', notwithstanding that the person in question does not fulfil the conditions as to residence or presence in Great Britain.[107] As no special rules apply in relation to the payment of oil-rig workers the liability to pay contributions of, or in respect of, a person employed on the continental shelf is thus exactly the same as it would be if he were employed ashore. Again it should be noted, however, that the regulations exempt the employees themselves from requirements made in the Act or regulations relating to the performance of any act forthwith, or within a specified time or following a certain event if this is rendered impossible by absence from Great Britain. Performance 'as soon as is reasonably practicable' will suffice.[108]

Oil-rig workers, and indeed those who are employed on pipe-line works, are thus entitled to all the benefits provided by the Act, subject, of course, to their meeting all the various rules and regulations, particularly those relating to contributions, and their qualifying in all other respects. Specific provision is made in the regulations to ensure that they or their families are not deprived of benefits under section 85 (5) (a) of the Act due to the worker's absence from Great Britain, providing that such absence is due to his being or having been in prescribed employment,[109] and again the provision that the performance 'as soon as is reasonably practicable' of those acts which must normally be done straight away or within a certain time if his absence makes meeting these requirements relating to benefit impossible is applicable.[110]

VIII. NOTIFICATION OF ACCIDENTS AND DEATHS

It has already been indicated that the manager of an offshore installation must report all casualties.[111] Under regulations made under the Mineral Workings (Offshore Installations) Act he must do this to the owner 'immediately', 'in the most expeditious manner practicable', giving brief particulars of the casualty, the position of the installation, the time of the accident and the identity of the person killed, lost or seriously injured.[112] In fact a description in writing also has to be made within three days.[113] The owner must in turn pass this information to the Secretary of State, and must also inform him if a person injured subsequently dies.[114]

In fact, if anyone does die (no matter how) on an installation, or is lost from an installation in circumstances in which it is reasonable to believe that he had died, or dies in or is lost from a lifeboat or raft or survival craft belonging to an installation, or if anyone is lost or dies in the neighbourhood of an installation in this way while connected with the installation, the manager of the installation is also under a duty to make a return of death to the owner of the installation of the death within ten days of its coming to his knowledge.[115] The owner then has the task of informing the Registrar General of Shipping and Seamen within a further ten days, and the duty of informing the dead person's next of kin, or, if he was not the employer of the dead person, the employer, who then presumably passes the information on himself.[116] This must be done within forty-eight hours at the outside.[117] In any event the installation manager should inform the local coroner, and this is usually taken to mean the coroner for the area in which the company has its land base, i.e. Great Yarmouth, Lowestoft, Aberdeen, etc.

The manager must also make a return relating to disabling accidents, injuries or diseases.[118] This is made every three months, and must contain brief particulars of the accident, injury or disease.[119] In relation to the serious accidents ('casualties') which the manager reports immediately, the regulations also require that no person shall disturb the place where it occurred or tamper with anything there without the permission of an inspector until either three days after the owner has given notice to the Secretary of State, or an inspector has concluded an investigation.[120] It is in this connection that the wide powers given to inspectors are mostly used.

IX. INQUIRIES AND INVESTIGATIONS

The members of the 'Sea Gem' Tribunal of Inquiry were so impressed with the generous contributions of all those who appeared before them and co-operated with the inquiry that they were moved to conclude that 'there can hardly ever have been an investigation better served by those who offered evidence'.[121] Indeed, it was described in the House of Commons as 'one of the most effective and powerful inquiries which had ever taken place'.[122]

It comes somewhat as a surprise, therefore, to learn that the Tribunal was entirely without statutory authority, had no power to compel the attendance of witnesses, and was not empowered to administer oaths.[123] The fact that the Minister lacked power to enforce an inquiry 'astonished us all when the "Sea Gem" went down', according to one MP,[124] though the gap itself springs from the formerly uncertain status of offshore installations. If registered as a ship there would have been power to hold an inquiry under the Merchant Shipping Acts, as there would had it been covered by the Factories Act. Not unnaturally, the 1971 Act sets out to fill this quite undeserved gap and to provide any future tribunal with the powers essential to its proper working.

The Schedule to the Mineral Workings (Offshore Installations) Act 1971 sets out some of the matters that may be covered by regulations made under section 6 (3) of the Act. Paragraph 10 of the Schedule provides that 'casualties or other accidents involving loss of life or danger to life' may be covered, and divides the possibilities into special reports by inspectors, and the holding of public inquiries. The wide powers of an inspector to hold an investigation on board the installation have already been dealt with, but in serious situations such as that in which the 'Transworld 61' and the 'Transocean III' found themselves, such an investigation may only be a preliminary to a proper inquiry.[125]

Under regulations the Secretary of State may appoint a competent person to conduct an inquiry where a casualty or other accident involving loss of life or danger to life which occurs on or in connection with the presence or operations of an offshore installation in a designated area.[126] The person appointed may be assisted by assessors, also appointed by the Secretary of State, and together they form a Court of Inquiry with power to hold an inquiry in public.[127] If the court is of the opinion that it is in the public interest that any part of the evidence or argument should be heard *in camera,* however, it may so direct.[128]

Such a court of inquiry is given full powers to carry out its work. It may employ experts to assist it,[129] and both the members and the experts have power to obtain access to and inspect any installation in a designated area or any part of such an installation, or any vessel which is, or has been, employed in connection with operations carried out from it.[130] They may enter any land or premises they consider necessary.[131] The court has full power to require the attendance of persons to give evidence or produce any document in his custody or under his control. It may allow persons before it to call witnesses to give evidence, and examine them. It may administer oaths and take evidence on oath and order any person before it to give a copy of any document produced at the inquiry to any other person also appearing at the inquiry.[132] Provision is made for a penalty of £200 on summary conviction for any person who fails without reasonable excuse to comply with any summons, citation or requirement of the court after having any expenses to which he is entitled tendered to him, and for impeding the court in the execution of its duty.[133]

Under the regulations the court must make a report to the Secretary of State stating fully the circumstances of the casualty or other accident with the opinion of the court as to its cause, 'together with any observations and recommendations which the court thinks fit to make with a view to the avoidance of similar occurrences in the future and the preservation of life'.[134] In addition, power is given to the court to direct that the costs of the inquiry shall be paid in whole or in part by any person who the court thinks contributed to the casualty or other accident necessitating the inquiry by reason of 'any act or default on his part or on the part of any agent or servant of his',[135] provided he has been given an opportunity throughout the hearing to play a full part in the proceedings.[136] This provision is of considerable importance, introducing as it does an element of civil punishment for an accident involving loss of life, or even danger to life. Any person found by the tribunal to be at fault in such circumstances might subsequently find himself liable to pay for the mistake up to three times; before the tribunal of inquiry, in a civil action for damages, and in a criminal trial if the act contravenes any of the regulations made under section 7 or any of the criminal sections of the Act. As the tribunal will undoubtedly be held first, this possibility should always be borne in mind.

Although no specific extension of the powers of coroners has been made in relation to deaths on the continental shelf, a coroner for the country into which a body is brought may also hold an inquest into death of that person no matter where it happened if he has reasonable cause to believe that the death was violent or unnatural or a sudden death the cause of which was unknown.[137]

In relation to Scotland, however, specific provision for the holding of inquiries into offshore accidents resulting in death has now been made in the Fatal Accidents and Sudden Deaths Inquiry (Scotland) Act 1976, which replaces a similarly titled Act of 1906 and the Fatal Accidents Inquiry (Scotland) Act 1895, both of which were extended to cover such accidents in 1975.[138] Under the 1976 Act the Procurator-Fiscal for the district in which it appears that an accident resulting in death has occurred to an employee in the course of his employment, or to an employer or self-employed person engaged in his occupation, is charged with investigating the circumstances of the death and applying to the Sheriff for the holding of an inquiry.[139] For the purposes of the Act a death which occurs in connection with the exploration of the sea bed or subsoil or the exploitation of their natural resources in the area of the UK continental shelf in which Scots law is applied is taken to have occurred in Scotland.[140]

The Procurator-Fiscal is placed in charge of all the inquiries, the citation of witnesses (who may be compelled to attend, but not to answer any questions which might tend to show them guilty of an offence), informing relatives and the employer of the time of the inquiry, and adducing evidence to the Sheriff.[141] At the conclusion of the evidence (which does not have to be corroborated to be believed), and any submissions, the Sheriff must then make a determination as to the time and place of the accident and death and their cause or causes, the reasonable precautions (if any) by which they might have been avoided, the defects (if any) in any system of working which contributed to them, and any other relevant facts.[142] This determination cannot be given in evidence in, or found, any subsequent proceedings relating to the accident or death.[143] Indeed, if an inquiry is held under the Mineral Workings (Offshore Installations) Act, an inquiry under the 1976 Act cannot be held at all unless the Lord Advocate so directs.[144]

The Petroleum and Submarine Pipe-lines Act also makes provision for the regulations to be made with respect to the notification of accidents connected with pipe-lines, and for the holding of public inquiries into such accidents.[145] Provision is made for the regulations to cover, *inter alia*, entry and inspection for such purposes, for summoning witnesses to give evidence or produce documents, and for the administration of oaths and the making of declarations.[146] In addition, the regulations may also cover the payment of the costs of the inquiry by the person appearing to the person holding the inquiry to be responsible in any degree for the accident in question whether by any act or default of his own or that of his servant or agent.[147] It is further specifically laid down that in the event of such an inquiry being held no inquiry shall be held under the Fatal Accidents and Sudden Deaths Inquiry (Scotland) Act unless the Lord Advocate so directs.[148]

In all, the employee on an offshore installation has the full range of legal remedies at his command in the event of an accident occurring in which he is injured in some way. The most important point is quite simply that the occupation is one in which accidents might occur more often than in many shore-based employments; the natural hazards are greater, and the equipment and the nature of the job are of such a nature as to be inherently more dangerous than those of the average employment. It is only through legal regulation of the kind outlined above that greater safety will be ensured, but it must be remembered that, in this case, the companies themselves are well aware of the cost of accidents to the drilling programmes, and this factor has ensured that a great deal of attention has been directed at the provision of as safe a system of work as is possible in the circumstances. The total elimination of danger is, however, impossible.

7 DISCIPLINE, DISMISSAL, AND THE CRIMINAL LAW

'Fall off that derrick, buster, and you're fired before you hit the deck.' (American foreman on an offshore rig. Quoted in the *Guardian,* 20 September 1973, p.13.)

Although in many ways central to the analysis of the differences between offshore workers and shore-based workers, the provisions relating to discipline can be thoroughly appreicated only against the safety requirements and the command structure operative in the industry. This is in fact an inevitable consequence of the legislation itself. As for many other workers, discipline and dismissal are intricately connected as the principal method of discipline has always been summary dismissal. As a result there are intermingled industrial and legal rules.

It is proposed in this part to deal with the specifically disciplinary provisions of the law first, and to add a few words on the wider application of the criminal law to offshore operations, then to consider the industrial discipline and the reasons for dismissal, and finally to look at the legal requirements relating to unfair dismissal and redundancy.

I. DISCIPLINE

The 'Sea Gem' Report did not only recommend a 'chain of command' for the offshore industry. It also demanded 'an accepted discipline'. Both were seen as being 'analogous to what has been established over the years by the practice of the Merchant Service fortified by the provision of the Merchant Shipping Acts'.[1] The Tribunal admitted that the former recommendation was 'suggested strongly' by the fact that: 'Although many of the perils from which those serving on drilling rigs must be protected are not in their essence maritime perils the 'Sea Gem' was lost in the character of a sinking ship. . .'.[2] Discipline was thus seen as an integral part of the command system.

Apart from this conclusion, no real justification is provided in the Report for these recommendations. Indeed the Tribunal appears to have condemned itself in this respect with the admission that:

The field over which the Inquiry ranged is so large. . .that several

> working parties able to dedicate a great deal of time to their studies would be needed if detailed recommendations were to be looked for. There is now a large number of oil drilling rigs of various nationalities, ownerships and types on location in the North Sea so that the generalisation could well be both inept and dangerous. It is unlikely that another drilling rig exactly like the 'Sea Gem' will ever be constructed so that the recommendations based upon what the Tribunal might think would have made the 'Sea Gem' a safer structure are little to the purpose.[3]

Nevertheless, the recommendations of the Report are generalisations from the 'Sea Gem'.

Superficially, certain comparisons can be made between the work situation of oil-rig workers and seamen. But no study has been undertaken by the Government to see if there are significant differences – a more difficult task than drawing comparisons. As a result it is not altogether surprising to find that the recommendations were taken at their face value, despite the warning given by the Tribunal. The various sections of the Mineral Workings (Offshore Installations) Act dealing with discipline, peculiar features of the legislation which distinguish it from that of other countries, are therefore not very different from some of the provisions of merchant shipping legislation, in particular the Merchant Shipping Act 1970. They take the attitude that discipline is an integral part of safety.

Not unlike those given to the master of a British ship, the powers given to the manager of an installation in the UK sector of the continental shelf are of a general nature. Section 5 (2) of the Act provides:

> Except as otherwise provided by this Act, the manager of an installation shall have in relation to it general responsibility for matters affecting safety, health or welfare or, where connected with safety, health or welfare, the maintenance of order and discipline, and for the discharge of that responsibility shall exercise authority over all persons in or about the installation: Provided that this subsection shall not extend to any matters for which another person is responsible as master, captain or person in charge of any vessel, aircraft or hovercraft.

The most important limitation on the manager's general authority is clearly the requirement that the matter must concern 'safety, health or welfare'. This was introduced at a late stage in the debates on the Bill as a result of representations made by opposition members with experience of the shipping industry,[4] but the result is by no means clear. The terms are nowhere defined in the legislation, and the range of matters which have been subsumed under these headings is enormous. Indeed they could be said to cover all aspects of life on board an installation,[5] though the effect should be to eliminate the duty of members of the crew to obey any capricious whim of the manager, such as fetching him a drink in his cabin, should that be outside the terms of their contracts of employment. But this later implied limitation is also important. The manager will also have the powers which are his by virtue of his position in the industrial hierarchy, powers accorded to him by the company and not by the Act, the ultimate sanction for which is the dismissal of the disobeying crew member.

The intermingling of contractual powers and those derived from the legislation is best illustrated in section 5 (3). This provides that: 'If a person subject to the authority of the manager of an offshore installation wilfully disobeys a lawful command given to him by the manager in exercise of that authority, he shall be liable on summary conviction to a fine not exceeding £50.' As the obligation to obey only arises when one is under the authority of the manager, it is also confined to the situation where the order concerns safety, health or welfare. But although this is the obvious intention of the legislation[6] the fact that the employee might be subject to the authority of the manager through his contract of employment is not considered in the Act. The contracts of employment of a great number of companies contain obedience clauses and clauses stipulating the grounds for which the employee will be dismissed, and workers are nearly always dismissed for breaking them. If this is the case, the scope of the grey area of contractual discipline outside section 5 (3) seems very limited.

On the other hand, the contrast with the special provisions relating to merchant seamen brings out the potential advantages of trade union representation for offshore workers. Comparison with the Merchant Shipping Act 1970 before its amendment in 1974 reveals that some of the safeguards incorporated into the analogous powers of the ship's master are absent from the legislation covering installations. Section 29 of the Merchant Shipping Act 1970 provided that a seaman may be liable to a £50 fine for disobeying a lawful command 'relating to or likely to affect the operation of the ship or its equivalent'. Although it seems clear that the breadth of the concepts used in section 5 (2) of the

Mineral Workings (Offshore Installations) Act is such as to render the equivalent limitation entirely void of meaning, no such safeguard found its way into the legislation for oil-rig workers, and the section remains even though the seamen have managed to secure the repeal of the equivalent section in the Merchant Shipping Act. In effect, the merchant shipping legislation, having spawned the provisions of the Mineral Workings (Offshore Installations) Act, has now passed it by, leaving it with the ancient concept of 'disobedience to a lawful command', again not really defined so that one could be sure of its extent. In consequence, little advice as to what constitutes a lawful command can be offered, except that it is most likely to be regarded by the courts as co-terminous with a reasonable instruction given to an employee in the course of his employment relevant to the obligations imposed upon him by his contract.

Further lack of limitation of the installation manager's authority, again clearly inspired by the 'Sea Gem' disaster, appears in section 5 (5). This provides that:

> Where at an offshore installation there is an emergency or apprehended emergency endangering the seaworthiness or stability of the installation or otherwise involving a risk of death or serious personal injury, the installation manager may take or require to be taken any such measures as are necessary or expedient to meet or avoid the emergency; and no regulation or condition having effect by virtue of this Act shall apply to prohibit or restrict the taking of any such measures by virtue of this section.

Although this section may well have been intended to protect the manager against liability should he be forced to throw valuable equipment overboard to preserve balance in an emergency, the result is to give him unlimited powers in the circumstances described. Again, an exact definition of an emergency is not provided and could not really be expected. In effect, the section would act as a justification for actions taken by the manager when questions of the reasonableness of his apprehension of an emergency of one kind or another might be relevant despite words which seem to indicate that simple apprehension, even without cause, would suffice.

In this respect, a contrast might be drawn with section 5 (6), which reads:

> If the installation manager has reasonable cause to believe that it is

> necessary or expedient for the purpose of securing the safety of an offshore installation or persons in or about it, or maintaining order and discipline among those persons, the installation manager may cause any of those persons to be put ashore in the United Kingdom; and where any of those persons has done or is about to do any act endangering or likely to endanger the safety of the installation or persons in or about it or the maintenance of order or discipline among those persons, or the installation manager with reasonable cause suspects him of having done or being about to do any such act, the installation manager may take or cause to be taken such other reasonable measures against him, by restraint of his persons or otherwise, as the installation manager thinks necessary or expedient. . .

The fact that this subsection gives to the installation manager all the powers he could possibly need to cope with any of the crew in an emergency, however defined, would appear to support the submission that the previous section is more concerned with property than people. It also reinforces the argument that safeguards might have been introduced to advantage in that subsection. Section 5 (6) itself is limited in application to circumstances in which the manager has 'reasonable cause to believe' that it is necessary or expedient that some measures be taken, or to circumstances in which the manager has 'reasonable cause to suspect' that action defined in the subsection has been, or is about to be, taken by the person in question. The measures taken, including imprisonment, must also be 'reasonable'. The test in each case is objective.

In favour of this section it must be mentioned that the equivalent section of the Merchant Shipping Act does not even require the master to act reasonably.[7] Indeed, offshore workers are further protected by section 5 (7) which limits the power to restrain to twenty-four hours unless the intention is to have the man flown off at the earliest opportunity and, in which case, the Secretary of State must be notified of the restraint (and the reasons for it) by the manager[8] within the twenty-four hour period or as soon as practicable. On the other hand, the whole of section 5, including these provisions, may be excluded in relation to any particular installation by the Secretary of State 'in such manner as he thinks fit',[9] and indeed any class or type of installation might be excluded by regulations. No such regulations have been made.

Three further points must be made to complete the picture of the statutory disciplinary measures provided for by the Mineral Workings (Offshore Installations) Act. Firstly, in the case of any offence alleged to have been committed in contravention of any of the parts of section 5, a defence is provided by section 9 (3) if the accused can prove –

(a) that he has used all due diligence to enforce the execution of this Act, and of any relevant regulation made under this Act, and
(b) that any contravention was committed without his consent, connivance or wilful default.

This subsection is clearly intended to provide the manager with a defence in the case of his breach of duty, for example, to inform the Secretary of State that he has restrained a member of the crew. It is not apposite to provide a defence to the disobedience of a 'lawful command' under section 5 (3).

It might be possible, though, for the manager to give a 'command' (though it is questionable whether it would be 'lawful') to a member of the crew to do something which is contrary to one of the regulations made under the Act, and which is nevertheless something relating to safety, health or welfare. The crew member might then refuse to carry out the instruction knowing of the existence of the relevant regulation. Could he then be successfully prosecuted under section 5 (3)? On the other hand, if he obeys the order of the manager, could he be prosecuted for breaching the regulations? This juxtaposition of the two questions does emphasise a dilemma unsolved by section 9 (3). The worker might be able to show that he was attempting to uphold the regulation, and he must therefore protest even if he later contravenes the regulation at the manager's request. In that case the breach of the regulation would not be 'wilful' but done under duress. If he simply refuses to act he seems to be outside section 9 (3) because, even though he was attempting to uphold the Act, the breach of one of the provisions, in this case section 5 (3), was most definitely wilful within the wording of both the relevant subsections. On the other hand, if although ultimately prepared to perform the request he protests and is then dismissed, whether or not the protest amounted to disobedience under section 5 (3), might be relevant in proceedings brought to determine that the dismissal was unfair. In all cases, therefore, even those involving breaching the regulations, it would seem advisable for the crew member to obey the manager's orders.

One further protection, and one which has general application to the criminal provisions of the Act, is added by section 10 (3) and (4) which require the consent of the Secretary of State (or various Law Officers[10]) before proceedings can be instituted for any offence under the Act. Though this is intended to cover any problems which might arise in the case of foreign nationals working on the British continental shelf,[11] neither of these subsections 'prevent the issue or execution of a warrant for the arrest of any person in respect of an offence, or the remanding in custody or on bail of any person charged with an offence'.[12] Discipline is therefore assured even though prosecution is not necessarily certain, a sentiment which receives further promulgation in section 8 (3) and section 12 (4) which declare that the Act applies 'to any individual whether or not he is a British subject'.

II. GENERAL CRIMINAL PROVISIONS

Most of the criminal provisions contained in the Mineral Workings (Offshore Installations) Act apply to the managers and owners of the rigs, or to the concession owners,[13] rather than to the persons working on board. This is supposedly also the intention of the regulations to be made under section 7 (2) of the Act,[14] although no such limitation appears in the Act itself. The subsection reads:

> Subject to subsection (3) below, regulations under this Act –
>
> (a) may provide for the creation of offences and for the punishment on summary conviction or on conviction on indictment, and
>
> (b) may afford, in respect of any description of offence created by the regulations, such defence, if any, as may be specified in the regulations.[15]

As with certain sections of the Act, the operation of the regulations may be excluded in certain particular cases, and for certain classes or types of offshore installation.[16] On the other hand all regulations made under the Act are to be contained in statutory instruments 'subject to annulment' in the Houses of Parliament,[17] a procedure that does not always lead to close scrutiny. Nevertheless, the penalties are limited. Section 7 (3) provides:

> The punishment for an offence created by regulations under this Act shall be –
>
> (a) on summary conviction a fine not exceeding £400,

(b) on conviction on indictment imprisonment for a term not exceeding two years, or a fine, or both,

but without prejudice to any further restriction on the punishments which can be awarded contained in the regulations, and without prejudice to the exclusion of proceedings on indictment by the regulations.

Although the penalties are not excessive if they are intended to apply only to the management side of the industry, there is no positive commitment to exclude the application of further, and more specific, disciplinary provisions to those who actually work on the rigs. If so, this is a further example of the futility of having a duty to consult with representatives of 'persons who will be affected by the regulations'[18] when there are as yet no unions with either extensive membership or recognition in the offshore industry.

In the Report stage of the progress of the Bill through the House of Commons an attempt was made to limit the power to create offences in the regulations to situations 'whereby the safety of persons on or near the installation, or of the installation itself, is endangered'.[19] The object was further limitation of the powers of the manager. The Government was not minded to accept this amendment, which was consequently defeated, because it thought that the power to create offences was already limited to those matters which could be the subject matter of regulations. These would be primarily related to safety, and some room to manoeuvre in relation to registration of installations, inquiries into casualties and accidents, and other matters relating to health and welfare matters was also needed.[20]

On the other hand, the Act does facilitate the prosecution of management in the case of the real blame for an incident being a personal one. Section 9 (1) provides:

> Where an offence under this Act which has been committed by a body corporate is proved to have been committed with the consent or connivance of, or to be attributable to any neglect on the part of a director, manager, secretary or other similar officer of the body corporate, or any person purporting to act in any such capacity, he, as well as the body corporate, shall be guilty of that offence and shall be liable to be proceeded against and punished accordingly.[21]

This formulation deals at one blow with several thorny problems of company law, such as piercing the corporate veil and the ostensible authority of officers. Furthermore, the spirit behind this intention to ensure that the offenders shall in all cases be apprehended is evinced by section 12 (4). This provides:

> It is hereby delcared that, notwithstanding that this Act may affect individuals or bodies corporate outside the United Kingdom, it applies to any individual whether or not he is a British subject, and to a body corporate whether or not incorporated under the law of any part of the United Kingdom.[22]

But although the application of the Act may well be just as important as far as foreign companies are concerned, enforcement of the provisions of the Act, like the availability of remedies to the individual worker, may well be more difficult.

III. INDUSTRIAL DISCIPLINE AND SUMMARY DISMISSAL

In an isolated workplace with high pressure to produce results, trouble of any kind cannot be tolerated by the management. The one sure way of dealing with an offender against the code of such a 'total' institution is to 'fly him off'.[23] The helicopter is the substitute for what might otherwise be a more lengthy disciplinary inquiry, centred round a more flexible code. In fact there is a great deal of consensus amongst the offshore operating companies as to which offences carry dismissal as the expressly stipulated punishment. These are regrettably not always made known to employees in accordance with the provisions of the Industrial Relations Code of Practice,[24] but many companies stipulate in the contract of employment that the employee may be summarily dismissed for a variety of reasons (indeed, contracts with one company say 'for any reason'), and others refer the employee to the company's general disciplinary code.

The offences which are detailed in the contracts can be grouped into four main categories. Firstly, there are those offences which are incompatible with membership of an enclosed, or 'total' community. Good examples are various sexual offences, and theft, of which several instances have occurred and dismissals followed as a result. Transworld also include improper behaviour – a rather wider term – and venereal disease amongst their reasons.

Secondly, there are offences which are related to safety, although these could in certain circumstances also be seen as offences against the system of work. Thus most companies will dismiss an employee for an unstipulated 'unsafe act'. If the rig permits alcohol, drunkenness is an offence, and if it does not, drink of any kind. A typical provision reads: 'The company reserves the right to terminate your employment without notice or payment in lieu of notice should he report for work under the influence of drugs, or drink, or become drunk or drugged whilst on duty.' Some other companies substitute 'or drink or take drugs whilst on duty' for the last phrase. Smoking in an area where it is prohibited falls into this category. It is a frequent reason for dismissal.

Offences against the system of work are of several kinds. Reading & Bates dismiss for failure to catch the outward flight for the beginning of a new duty tour. Being absent from work without permission is a frequent reason, though rather more comprehensive. Transworld will dismiss for 'non-compliance with regulations' and also 'wilfully inflicted self-injury'. Many companies stipulate that dismissal will follow any injury or sickness that renders the worker incapable of performing his duties, even temporarily. Incompetence, of course, also falls under this category, as does insubordination and failure to obey the instructions of the toolpusher or immediate supervisor, or to carry out duties assigned by those superiors.

Finally there are acts which can be seen as offences against the interests of the company. Thus breach of security or industrial espionage will meet with instant dismissal. Many will dismiss employees should they have lied in any application form or document relating to their abilities. Many companies also write in specific clauses which relate to divulging of information about the company's affairs to anyone not employed by the company, or divulging information about a well drilling. Until recently, British Petroleum had a clause which stated that dismissal would follow if the employee 'wilfully neglects the company's interest'.

One special offence merits particular comment. To work on an offshore installation the workers all have to have passports as the rig is technically outside the country. The oil companies have a special relationship with the customs and excise so as to obviate the need for constant passage through the customs for little or no reason. Indeed, there is little to be brought back from the rig which is worthy of declaration except cigarettes bought from the offshore caterers. Obviously this special relationship would be put in jeopardy by evasion of the restrictions perpetrated by any of the employees and the special dispensation and arrangements are extremely valuable to the companies because of the amount of time saved. Customs evasion is therefore clearly defined as meriting instant dismissal. The same could be said of contravention of any of the immigration regulations. In both cases it is interesting that the companies accord these offences the status of 'serious offences'.

It should be noted, however, that many of the 'offences' are couched in extremely vague terms. 'Misconduct' is a case in point. The Industrial Relations Code of Practice states that in cases of misconduct 'the first step should (normally) be an oral warning or, in the case of more serious misconduct, a written warning setting out the circumstances'.[25] It also suggests that 'no employee should be dismissed for a first breach of discipline except in the case of gross misconduct'.[26] All the evidence is, however, that discipline on offshore oil-rigs is rigid, and in many cases harsh and immediate.

Given this wide-ranging industrial control exercisable by the management over the crew of an offshore installation, one must again ask the question: why was it thought necessary to supplement it with legal powers analogous to those possessed by the master of a ship under the Merchant Shipping Act? This is especially relevant in that the justification for the powers of ships' masters was itself being criticised,[27] and has subsequently been amended. The answer lies in the attitude of the 'Sea Gem' Tribunal towards installations, and the analogy drawn in the Report between a sinking ship and a sinking installation.

But underlying the acceptance by the Government of this comparison seems to be an equally unquestioned acceptance of certain remarks of the Pearson Court of Inquiry into certain matters concerning the shipping industry as valid for all maritime institutions. The Pearson Report came to the conclusion that:

> Discipline is necessary in the interests of safety, in order to secure prompt and co-ordinated action by the crew on any occasion of emergency. . .Discipline is necessary in the interests of efficiency. . . Also, discipline is necessary for preserving law and order in the confined and inescapable conditions, because the ship is a 'total' institution, in which the seaman works, eats, sleeps and spends his leisure time in the same limited area and with approximately the same people as his workmates, companions or neighbours. Good discipline must be maintained somehow.[28]

The members of the Court of Inquiry thought that some parts of the system could only be called into question 'if all. . .were ready to understand, and adapt themselves to, the operational requirements of maritime business and to appreicate the special need for restraint and moderation and self-discipline in the conditions of working and living at sea'. But many are only temporary and casual workers and therefore 'it cannot be expected that all of them will always be adult and responsible and moderate in their behaviour'.[29]

The application of Erving Goffman's general theory of 'total' institutions, which he describes as places of residence and work 'where a large number of like-situated individuals, cut off from the wider society for an appreciable period of time, together lead an enclosed, formally administered round of life',[30] has been readily accepted in relation to ships by a number of writers.[31] In this truncated form it would also appear to be an acceptable shorthand description of the environment provided by an offshore oil-rig. It should be noted, however, that even as a sociological model it has serious weaknesses, in particular the inability to weight differentially the force of any particular characteristic which may be exhibited by a large number of different organisations. The model is thus based upon the false premise that institutions identical with regard to their structure will impinge upon their inmates in identical ways regardless of differences in their backgrounds and the purposes of the institution itself.

Although this is not really the place to consider the various merits and demerits of Goffman's model, it is legitimate to exhibit concern over the uses to which it has been put, especially if translated into legal form. In this respect the Mineral Workings (Offshore Installations) Act provides another example of the misuse of what is primarily a descriptive concept in a prescriptive manner. The classification of the place of work as a 'total' institution would appear to have been taken to justify of itself the imposition of a special disciplinary system in supposedly isolated offshore communities. But this is to put the cart before the horse. The disciplinary system might aid the descriptive classification of the ship, and, indeed, of the offshore oil-rig, as a 'total' institution, remembering that Goffman denied that all total institutions possessed the same attributes.[32] But placing the institution into such a classification does not mean that a strict disciplinary system must therefore automatically be imposed within it.

Such criticism is forceful enough when made in relation to ships. It is even more appropriate in connection with offshore oil-rigs. Although the provisions of the Mineral Workings (Offshore Installations) Act do not go as far as the provisions of the Merchant Shipping Act 1970, in many respects, including, for example, on-the-spot fining, the circumstances of the employee in the offshore oil industry are not directly comparable with those of the seaman. There are great differences in the work and the life that goes with it. The recruitment, training, and background of the workers are not at all similar. The system of industrial relations that exists in the industries, both between companies and workers, in general and in the individual workplaces, do not bear comparison to the extent that one could consider them identical in very many ways. Ships travel and have a life that goes with travelling. Offshore installations are much more like offshore factories. Dissidents can be removed very quickly.

Without the further investigation and consideration demanded by the 'Sea Gem' Inquiry, the automatic application of the rationale of the Pearson Report to installations should not, therefore, have gone unchallenged. Discipline is obviously necessary on board. But does the absence of legal reinforcement of the industrial powers of management mean that there will be anarchy and chaos, especially in an emergency? Actual events on board the 'Sea Gem' at the time of its collapse do not point in this direction.[33] Before the passage of the Act there was no incidence of serious trouble on board any offshore installation that might now meet with the force of law which was not solved at the time with the use of ordinary industrial discipline.

If this is the case one can legitimately pose the question: why was it thought necessary to add further statutory powers in an area where they are, to all intents and purposes, superfluous to the needs of the industry, given the perfectly adequate powers possessed by the manager of an installation under the contract of employment? It must be remembered that Pearson was of the opinion that industrial disruption cannot be tolerated at sea,[34] and with the growing importance of the offshore oil and gas industry to the nation's economy, consideration of the force of a similar but unarticulated motive in the Mineral Workings (Offshore Installations) Act cannot be ruled out. Certainly one of the results of the provisions is to greatly restrict the ability of workers on the rig to exercise their industrial power, and even if there may be merits in that fact, the result is to force dissatisfaction ashore where it appears to be dissipated in the form of labour turnover. In such circumstances the presence of unnecessary and unused criminal sanctions would appear to be as reinforcing of resentment and potentially bad industrial relations as they are of good discipline.

IV. UNFAIR DISMISSAL AND REDUNDANCY

In relation to those statutory rights given to employees in relation to dismissal, the recent extension of the legislation under section 127 of the Employment Protection Act is of paramount importance. Now this has been done the application of the unfair dismissal provisions of the Trade Union and Labour Relations Act 1974, the provisions of the Redundancy Payments Act 1965, and those parts of the Employment Protection Act itself which are relevant in this connection, will give rise to fewer problems.

In relation to the right not to be unfairly dismissed, it will be recalled that the relevant provisions of the Trade Union and Labour Relations Act (and those improvements and additions made to it by the Employment Protection Act and the Sex Discrimination Act): '. . .do not apply to any employment where under his contract of employment the employee ordinarily works outside Great Britain.'[35] This formulation had the effect of taking those workers employed as oil-rig workers out of the scope of the provisions, and indeed at least two decisions of industrial tribunals held themselves lacking in jurisdiction to hear a claim by an oil-rig worker for precisely this reason.[36] At the same time it continued to give protection for those who were employed in the industry on shore but performed work on the continental shelf from time to time. In addition, those workers employed as merchant seamen aboard the supply boats were protected unless the boat was registered outside Great Britain (in other words the boat operated from another country), or the worker concerned was not ordinarily resident in Great Britain.[37]

The Employment Protection (Offshore Employment) Order 1976 took care of all these anomalies by simply redefining 'Great Britain' to include territorial waters (although this was probably not strictly necessary) and waters in any designated area other than one in which the law of Northern Ireland is applied. In addition, the Trade Union and Labour Relations Act is applied as though paragraph 9 (3) of Schedule 1 were omitted, thus including within the provisions those who are in effect seamen working on installations classified as ships but which remain within the boundaries of the designated areas.

A similar solution is adopted by the Order to the problem posed by section 17 of the Redundancy Payments Act 1965 which provides that:

(1) An employee shall not be entitled to a redundancy payment if on the relevant date he is outside Great Britain, unless under his contract of employment he ordinarily worked in Great Britain.

(2) An employee who under his contract of employment ordinarily works outside Great Britain shall not be entitled to a redundancy payment unless on the relevant day he is in

Great Britain in accordance with instructions given to him by his employer.

Apart from the question of the interpretation of the word 'ordinarily', this section posed somewhat different problems. Section 17 (1), by concentrating on the employees ordinarily working *in* Great Britain, appeared to make coverage of professional employees and others who normally worked ashore but visited the rig from time to time somewhat easier. In addition, an oil-rig worker who was dismissed by reason of redundancy while ashore might have been covered under section 17 (2).[38] This was not a question which received attention by the tribunals or the courts. Although BP had several redundancy cases brought against them, in none was the argument raised that the tribunal might not have jurisdiction.

Of course, the principal reason why the Act will not be of much relevance is the length of the qualifying period.[39] Many companies dismiss their roustabouts and semi-skilled workers if they move the rig from one sector of the North Sea to another for a prolonged period of time so that it would be closer to another country where workers of that country can be hired. Some will also have to dismiss their employees if drilling stops for the winter. Although senior personnel may be retained in either case, it is unlikely in the case of the other members of the crew. Few claims have thus been presented, which might also be taken as further evidence of the fact that most of the unskilled members of the crew of offshore oil-rigs view their jobs as casual ones, as indeed they frequently are.

With the recent extension of the Employment Protection Act itself, the various provisions of the Act relating to the notification of trade union officials and the Secretary of State in the event of a large number of men being dismissed together[40] have now become relevant in connection with dismissals because of the winter weather. But the possibility of complaints to an industrial tribunal in relation to those provisions, and the question of protective awards, will only arise where the employer has recognised a trade union or a recommendation for recognition has been made by the Advisory, Conciliation and Arbitration Service.[41] The extension also confers on oil-rig workers the right to receive a written statement of the reasons for their dismissal provided the worker in question had been employed for at least twenty-six weeks.[42]

There are two further potential problems relating to dismissal which are relevant to employment on an offshore oil-rig. Firstly, the computation of the period of continuous employment, now the various Acts listed in section 127 (1) of the Employment Protection Act 1975 have been extended to employment on the continental shelf, will pose problems because of the weeks the employee spends ashore. Some companies have in the past not regarded themselves as the employers of the worker for this period, although most, and in particular those who pay their workers monthly salaries, regard the worker as still being their employee. In which case the fact that he does not actually work will not interrupt the continuity of the employment, in accordance with paragraph 5 (1) (c) of the Contracts of Employment Act 1972.

Secondly, in all cases of dismissal of a loaned or borrowed employee the proper procedure to avoid problems arising under the contract of employment is to return the worker formally to his employer, in some cases the parent company or another subsidiary of the same parent, with a request that he be dismissed by them. This practice is essential, and indeed is normally followed, in the case of companies with a complicated structure such as Shell, and many of the American companies operating in the North Sea.

Finally, in the event of unemployment following dismissal the offshore worker should be eligible for unemployment benefit, subject to the normal rules, under the extended provisions of the Social Security Act 1975.

8 CONCLUSIONS

> 'What the lawyer can do and the legislator can do remains important, far more important is the work of the engineer and the scientist, the practical economist, and the creative organiser. . .I regard the law as a secondary force in human affairs, and especially in labour relations.' (Otto Kahn-Freund, *Labour and the Law,* Stevens & Sons, London, 1972, p.3.)

For the most part the problems of labour law as they relate to workers on the continental shelf in the North Sea have two causes. The first is the formula by which it was decreed that law should apply to operations on the shelf. The second is the type and style of the companies that make up the offshore industry, and the pressures, political and economic, under which they operate. There are other causes too. The very nature of the work and the place where it has to be performed provide their own sources of difficulty.

Throughout the legal provisions relating to offshore work runs the distinction to be drawn between those which automatically apply because the installation is within territorial waters, and those whose application is dependent on the tortuous provisions of section 3 of the Continental Shelf Act or specific extension by a suddenly mindful legislature. How much more straightforward it might be if the legislation adopted a more simple approach such as that taken in countries like Norway, Denmark or even France or the USA and directly applied the law of England, Scotland and Northern Ireland to installations and activities in the relevant parts of the UK sector of the continental shelf relating to the exploration of the sea bed and subsoil and the exploitation of their natural resources. In fact, of course, such an approach would not end the difficulties of regulating offshore activities nor the problems of relating labour law to offshore work. Apart from the outstanding example of the Race Relations Act 1968, the present approach, especially since the passage of the Employment Protection Act 1975, enables the majority of the legal problems to be taken care of in such a manner that certain allowances for the particular difficulties of working offshore may even be made as well.

But the very fact that regulations have to be made at all is a source of weakness, for it encumbers the government department charged most particularly with carrying out other potentially conflicting policies relating to the development of the resources in question with the task of imposing what are inevitably limitations on companies' freedom of action at the same time. It is here that the political and economic factors come into play and expose the potential weakness of any group of workers in a highly capital intensive industry, and particularly in one dominated by huge multi-nationals whose resources and expertise are desperately needed by a country seeking to rid itself of a crippling balance of payments deficit largely accounted for by the cost of importing the very product the companies are being asked to develop. As these same concerns are also involved in the production of imported oil too, and have in some cases huge capital sums tied up in the transportation system at the same time, it is not difficult to see that unless the benefits of involvement in a particular area match their expectations, within certain margins of course, the temptation to transfer operations elsewhere, or to make rather less effort than the government might like, cannot be far away. Hence not only do matters of politics, such as the rate of taxation to be levied on oil companies in respect of their offshore operations or the degree of public participation that is to be forced upon the multi-nationals because 'after all it is our oil', contribute to the solution to questions of job opportunities and job security, but so does the operation of laws which give labour an effective say in the terms and conditions of employment and the state and operation of the plant. The failure to extend such provisions in the legislation itself places their effective implementation, realistically speaking, on the bargaining agenda, and it should be noted that many sets of regulations have yet to be made, most of which touch on labour matters.

The same is true of the manner of enforcement of laudable regulations which attempt to regulate and improve working practices from the point of view of their safety. The maintenance of the safety inspectorate in the Department of Energy is of itself threatening to effective supervision of North Sea operations, even if it makes the most sense from an administrative point of view. Not until the Health and Safety Executive is entrusted with enforcing the regulations will this no longer apply, even if other problems, such as the inability of any inspectorate to make surprise visits to a rig out in the North Sea when they have to be made by helicopter or boat, and the enormous difficulties of constantly adapting standards to rapidly changing advanced technology, are inherent to some degree in the nature of the operations.

The nature of the operations does, it seems, provide justification for a number of other measures which would not go unnoticed ashore. The special disciplinary provisions are a case in point and are a peculiar feature of the UK legislation. In many respects it is strange that they should have been introduced into the legislation in an era when equivalent provisions in allied legislation were being toned down, but this is to some extent explained by the change in the political character of the Government between the two Acts of Parliament in question. Overall, however, the Mineral Workings (Offshore Installations) Act would appear to have the potential to regulate all manner of continental shelf operations in such a way as to give satisfaction to both sides of industry, and indeed to the public, subject to reservations just expressed about the regulatory process. There should be little doubt of the value of special legislation for special groups after examination of this particular example, provided that such legislation is consistent and not discriminatory, acceptable and not unnecessary. The main problem with such criteria in this industry at least has been the absence of trade unions to represent the point of view of those for whose benefit the legislation was passed in the first place, and here it is again a matter of regret that section 127 of the Employment Protection Act was only enacted in 1975. Had a similar result been achieved by the Continental Shelf Act the process of consultation might have had more meaning.

Indeed it is section 127 of the Employment Protection Act which places the industry at somewhat of a cross-roads as far as labour relations are concerned, with important repercussions lying ahead whatever the reaction. In many respects this is true of all the legislation passed by Parliament in 1975 which has reference to offshore oil, and the industry now has to decide how to live with a number of measures which offend many of its traditional notions. The taxation and participation provisions strike at the purpose of heavy investment in such risky operations, although they might have been foreseen. The encroachment of union power and the potential limitations on the power of dismissal run counter to the mode of operation of many of the companies from overseas operating rigs in the North Sea. It will be interesting to see how these measures relate to future developments.

A number of other conclusions may also be drawn from this study. In particular the process of legislative reform itself is revealed as primarily one of response to specific events, at least in relation to ostensibly uncontentious matters such as safety, widely regarded as totally lacking political implications. The legislation which forms the backbone of the system of employment regulation on the continental shelf springs almost directly from the work of a tribunal which by its own admission made very little inquiry into the subject matter subsequently covered by the Act or the regulations to be made under it. And yet its findings were accepted without question, and the implications of the provisions which Parliament might have reviewed were never thoroughly debated. Although it was recognised that control over safety matters through the licences was inadequate, no question was raised as to the adequacy of the method by which other laws relating to employment offshore were applied. The same Parliament that instigated the right not to be unfairly dismissed excluded workers on the continental shelf from the scope of those provisions. Just as the legislative process is one which constantly strives to keep pace with industrial and economic development, so it is one which must to a large extent rely on what it is told.

From the point of view of labour law the study also indicates the continuing importance of the individual contract of employment. Traditionally labour law as a discipline has been divided into individual and collective aspects, the one concentrating on the contract of employment, and the other concentrating on the more modern trade union. It is commonly thought that collective agreements have all but rendered the individual contract insignificant because of the scope of their coverage, and it is to the collective agreement that one naturally turns for the industrial rules that make sense of the legal provisions. But the offshore industry has no such collective terms, and the worker is employed under a contract he supposedly bargains for himself. It is such groups that serve as a reminder that there is a danger in assuming so much as to place rights in the hands of trade unions that are unavailable to individuals in such a way that not only are the individuals disadvantaged by not being members of unions, but are denied the opportunity to press legitimate grievances against their employer where unions have not as yet gained a foothold. Further, there is the danger that labour law will assume that all workers are covered by trade unions and concentrate unduly on the collective aspects of the employment relationship. For small but important groups such as those who work offshore such a development would be a mistake.

Finally, of course, it is hoped that the study serves to enlarge the range of measures that are considered under the rubric of labour law as well as filling a gap in the considerations of international lawyers. There is no such thing as the average man, nor the average worker, nor the average company or industry. Though the very nature of the employment relationship itself remains the same, the features of the work situation differ widely between industry, occupation, and indeed between companies. Through the greater appreciation of their differences may come a greater understanding of their similarities.

NOTES

CHAPTER 1

1. For rather general accounts of offshore drilling in the North Sea see Bryan Cooper and T.F. Gaskell, *North Sea Oil – The Great Gamble,* Heinemann, London, 1966; Peter Hinde, *Fortune in the North Sea,* Fowlis, London, 1966; and Clive Callow, *Power from the Sea,* Victor Gollancz, London, 1973.
2. See P.R. Odell, *Oil and World Power,* 2nd edn, Penguin, London, 1972, for a detailed analysis. On the oil industry generally see J.E. Hartshorn, *Oil Companies and Governments,* 2nd edn, Faber & Faber, London, 1967; C. Tugendhat, *Oil: The Biggest Business,* Eyre & Spottiswood, 1968.
3. For the history of oil prospecting and drilling on land in the United Kingdom, see: R.G.W. Brunstrom, 'Indigenous Petroleum and Natural Gas in Britain', in *The Petroleum Industry in the U.K.,* ed. Peter Hepple, London, 1966. There has in fact been a revival of onshore exploration in the UK. For example, some fourteen new onshore exploration licences were granted to Shell (UK) Ltd by the Department of Energy in 1975.
4. Later developments confirmed the field as the world's second largest. It was estimated at the time to be about 1,100 billion cu. ft. *(World Petroleum,* Mar. 1964, p.29) and was expected to yield about 30 billion cu. metres annually *(World Petroleum,* June 1964, p.88).
5. *World Petroleum,* June 1963, p.358.
6. *World Petroleum,* Nov.1964, p.36. This number had risen to 50 in 1964.
7. *World Petroleum,* Mar.1963, p.70.
8. *World Petroleum,* May 1963, p.56.
9. *World Petroleum,* Dec. 1963, p.8.
10. In fact gas was struck in June in a field 30 miles north of Juist Island, but found to be mainly nitrogen. See Richard Young, 'Offshore Claims and Problems in the North Sea', (1969) 59, *Am.J.Int.L.* 505, 1965.
11. See M.J. Wells, 'First Round of North Sea Drilling Starts', *World Petroleum,* Aug.1965, p.37; M.J. Wells, 'Hard Lessons Learned in the North Sea', *World Petroleum,* Mar.1966, p.23; M.J. Wells, 'North Sea Gas will Supply most U.K. Gas Needs', *World Petroleum,* July 1966, p.28; M.J. Wells, 'Operators are moving in around North Sea Gas Discoveries', *World Petroleum,* Sept.1966, p.44; Ray Gibson, 'The World's Drilling Industry', *World Petroleum,* Nov. 1966, p.30.
12. Discussed below.
13. This is also discussed further in Chapter 2.
14. See the Continental Shelf Act 1964, s.9.
15. See M.J. Wells, 'North Sea Future – A Question Mark', *World Petroleum,* Apr. 1967, p.180; M.J. Wells, 'U.K. North Sea Activities Still Slow due to Price Negotiations', *World Petroleum,* Sept.1967, p.25; G. Polangi, *What Price North Sea Gas?,* Hobart Paper No.38, Institute of Economic Affairs, London, 1967. See too the National Board for Prices and Incomes, *Report No.57, Gas Prices (First Report),* Cmnd. 3567, HMSO, London, 1968; id., *Report No.102, Gas Prices (Second Report),* Cmnd. 3974, HMSO, London, 1969; Graham L. Reid and Kevin Allen, *Nationalised Industries* (Penguin, 1970.), ch.3, esp. pp.63-81; G.L. Reid, 'The Gas Industry', ch.3 of G.L. Reid, Kevin Allen and D.J. Harris, *The Nationalised Fuel Industries,* Heinemann, London, 1973.

16. See M.J. Wells, 'Backdoor Nationalisation Proposed for North Sea Gas', *World Petroleum,* Sept.1967, p.14, and *North Sea Gas,* The Labour Party, 1967.
17. E.g. Texaco, Amoco, Conoco, Gulf Oil and Esso.
18. In 1970 the UK produced 26,286 Gigawatt hours of electricity by nuclear means. (1 Gigawatt = 1,000,000 kilowatts.) The USA produced 23,574, and the EEC countries 15,342. In 1971 the UK produced 27,642 (plus 5 per cent), the USA 39,494 (plus 67 per cent), and the EEC 18,918 (up by 23 per cent). See *Petroleum Press Service,* Sept.1972, p.346. According to the Department of Energy, the UK position has since remained roughly stationary. See *U.K. Energy Statistics* 1975, Aug.1975. For the development of the nuclear power industry, see W.G. Jensen, *Nuclear Power,* Henley-on-Thames, 1969. The decision in favour of the steam generating heavy water reactor (SGHWR) was announced on 10 July 1974, the same day the Labour Government announced its policies on increased public participation in the benefits of North Sea oil. See *H.C. Debates,* vol.876, col.1357 *et seq.* Cf. Harry Street and F.R. Frame, *Law Relating to Nuclear Energy,* Butterworths, London, 1966.
19. See R.M. Pegrum and D.R. Whitbread, 'U.K. and Ireland Offshore Activity on Big Upswing', *World Petroleum,* Mar.1971, p.30.
20. See B.C. Hague, 'North Sea Pace Picks Up Momentum', *World Petroleum,* Oct.1971, p.26.
21. Gas from the 'Frigg' and 'Brent' fields alone are expected together to supply over one-third of the country's gas needs by 1980.
22. By the Gas Council/Amoco group.
23. By British Petroleum. See British Petroleum *Press Statement,* 19 Oct.1970. It was confirmed as commercial in Dec.1972, see British Petroleum Information Department press hand-out, Apr.1972. See too 'North Sea also a Major New Oil Region', *World Petroleum,* Oct.1971, p.36.
24. See Peter Gillman and George Rosel, 'The Oil Rush', *Sunday Times,* 28 Apr.1974.
25. The Organisation of Petroleum Exporting Countries.
26. See Department of Energy, *United Kingdom Offshore Oil and Gas Policy,* Cmnd.5696, HMSO, London, 11 July 1974. It has since risen further to nearly $12.50 per barrel.
27. See *Petroleum Press Service,* Feb.1972, p.66 and p.99.
28. See *Petroleum Press Service,* Sept.1972, p.319.
29. For full details of finds see the *Study of potential benefits to British industry from offshore oil and gas discoveries,* International Management and Engineering Group of Great Britain Ltd., HMSO, London, 1973. (Referred to from now on as the 'IMEG Report'), Tables 1, 2 and 3; and the Department of Energy, *Development of the oil and gas resources of the United Kingdom,* HMSO, London, Apr.1975.
30. See *Production and Reserves of Oil and Gas on the United Kingdom Continental Shelf,* Report to Parliament by the Minister for Industry, HMSO, London, 1973, p.1. 'North Sea's Latest Estimates', *Petroleum Press Service,* 1973, p.204. See too the *Scottish Economic Bulletin: North Sea Oil,* HMSO, London, 1973.
31. IMEG Report. For a more overall view see *Oil: The Present Situation and Future Prospects,* OECD, Paris, 1973.
32. See the *Guardian,* 4 Jan.1974.
33. See Department of Energy, *Production and Reserves of Oil and Gas in the United Kingdom,* A Report to Parliament by the Secretary of State for Energy, May 1974, HMSO, London, 1974. This figure rose to 13 by Sept.1975, and is expected to go on rising.

34. *Petroleum Times,* 27 July 1974, p.4. Now confirmed as 'substantial'; *The Times,* 4 Sept.1975, p.17.
35. See the *Guardian,* 25 Oct.1974; *Petroleum Times,* 1 Nov.1974, p.4.
36. *The Times,* 19 Feb.1975, p.19; *Petroleum Times,* 21 Feb.1975, p.4. See too *Petroleum Economist,* Mar.1975, p.93.
37. See *The Times,* 25 Feb.1975, p.17.
38. See *The Times,* 6 May 1975, p.17.
39. *The Times,* 3 Sept.1975, p.17; *The Oil and Gas Journal,* 8 Sept.1975, vol.73, no.36, p.36; and *The Times,* 19 Jan.1976, p.13.
40. See *The Times,* 10 Apr.1975, p.19; *The Times,* 8 July 1975, p.17.
41. *The Times,* 1 Mar.1975, p.17, and *The Times,* 8 Apr.1975, p.21.
42. See *The Times,* 15 Oct.1975, p.15; *The Times,* 30 Aug.1975, p.15; *The Times,* 10 Sept.1975, p.21; *The Times,* 4 Sept.1975, p.17; *The Times,* 5 Sept.1975, p.15. BP, in partnership with the National Iranian Oil Co., also announced a discovery in Oct.1975 (see *The Times,* 28 Oct.1975, p.15) which will probably be developed because of its proximity to the 'Piper' field.
43. *Production and Reserves of Oil and Gas, etc.,* 1974, p.14.
44. *Development of the oil and gas resources of the U.K.,* 1975, pp.14-15. This upgrading of the reserves has taken place even against BP's downgrading of the possible size of the 'Ninian' field. See *The Times,* 3 Mar.1975, p.17, and *The Times,* 17 July 1975, p.17.
45. See *Petroleum Intelligence Weekly,* vol.14, no.16, 21 Apr.1975, p.9. In this context it is perhaps worth mentioning that in 1972 the total achievement, according to *Petroleum Press Service,* added up to 'less than one moderate sized Middle East Field', 1972, p.99. The estimated resources should be compared to the Middle East's estimated reserves of 340 billion barrels. As for the future of oil discoveries see H.R. Warman *et al.* in *The Geographical Journal,* vol.138, p.287 (Comment in *New Society,* 23 Nov. 1972, p.463); P.R. Odell, 'The Future of Oil: A. Rejoinder', *The Geographical Journal,* vol.139, p.436.
46. *Development of the oil and gas resources of the U.K.,* 1975, p.16.
47. Ibid., p.15. See too *The Times,* 15 Apr.1975, p.17.
48. *The Times,* 18 June 1975, pp.1 and 27.
49. *The Times,* 4 Nov.1975, pp.1 and 2.
50. See *Petroleum Intelligence Weekly,* vol.14. no.51, 15 Dec.1975, p.10; *The Times,* 15 Feb.1976, p.19
51. See *The Times,* 8 Jan.1976, p.16.
52. *The Times,* 30 July 1975, p.19.
53. See *The Times,* 21 Feb.1975, p.18, and *The Times,* 9 July 1975, p.17.
54. See *Development of the oil and gas resources of the U.K.,* 1975, p.3; cf. *The Oil and Gas Journal,* vol.72, no.21, 3 June 1974, p.103.
55. *Petroleum Intelligence Weekly,* vol.14, no.32, 11 Aug.1975, p.5.
56. See 'Texas of the North', *The Economist,* 29 Jan.1972, p.69. For further details of estimated effect of the discoveries on Scotland see B.C. Hague, 'Scotland and Offshore Oil', *World Petroleum,* Mar.1972, p.19; 'Scotland's Opportunity', *Petroleum Press Service,* Apr.1972, p.140. For more recent criticism see the *Report from the Select Committee on Science and Technology: Offshore Engineering,* Session 1974, H.C. Paper 313 of 1974, HMSO, London, 1974.
57. See 'North Sea Fact and Fiction', *Petroleum Press Service,* 1972, p.278.
58. See *Petroleum Press Service,* July 1972, p.260.
59. See *Petroleum Press Service,* 1972, p.278.
60. See 'Financing Future Growth', *Petroleum Press Service,* Aug.1972, p.291.
61. See 'North Sea Development Costs', *Petroleum Press Service,* 1972, p.121.

62. *The Oil and Gas Journal,* vol.73, no.26, 30 June 1975, p.74.
63. See Dr Jesse Wyllie, President of Gulf Oil, Eastern Hemisphere, in *The Sunday Times,* 22 Sept.1974.
64. *Petroleum Intelligence Weekly,* vol.14, no.18, 12 May 1975, p.12.
65. *Petroleum Intelligence Weekly,* vol.14, no.29, 21 July 1975, p.12.
66. See *The Oil and Gas Journal,* 13 Oct.1975, Newsletter; *Petroleum Intelligence Weekly,* vol.14, no.32, 11 Aug.1975, p.5.
67. See *Petroleum Press Service,* Sept.1972, p.342. The effect of oil on the share prices of British companies was reviewed by Christopher Stobart in 'Who's Who in the North Sea', *Investor's Chronicle & S.E.G.,* 31 Mar.1972, p.1042.
68. *The Times 1000,* 1972/3, Times Newspapers, London, 1972. See too *Labour Research,* 61, no.10, Oct.1972, p.224.
69. See 'Profitless Prosperity', *Petroleum Press Service,* Sept.1972, p.318.
70. *The Times 1000,* 1974/5, Times Newspapers, London, 1974.
71. BP made £295.5 million net profits in the first three months of 1974 alone. See the *Guardian,* 31 May 1974.
72. Esso, Texaco, Mobil, and Ultramar all rose. Only Amoco lost money.
73. See *Petroleum Times,* 21 Mar.1975. BP's net income rose from £329 million to £487.4 million, of which £175 million was 'non-recurring stock profit', i.e. profit from the oil crisis. A breakdown of the figures for all major companies can be found in *Petroleum Economist,* May 1974, p.170.
74. Indeed the profits of BP, Esso and Shell have all sagged. See *Petroleum Intelligence Weekly,* vol.14, no.36, 8 Sept.1975, p.6; vol.14, no.41, 20 Oct. 1975, p.12; vol.14, no.46, 10 Nov.1975, p.6; vol.14, no.49, 8 Dec.1975, p.9.
75. For example, in the takeover by the Marathon Company of the ill-fated Upper Clyde Shipyards. See DoTI *Press Notice,* 15 Mar.1972, ref.no.241.
76. See the National Coal Board (Additional Powers) Act 1966, c.47, esp. s.1 (1) (a), and the Gas Act 1965, c.36, esp. s.1 (2).
77. Industrial Development Act 1966, c.34, s.1 (1) and (4).
78. Explained below at page 46.
79. Industrial Development Act 1966, s.1 (2) (b) and (4).
80. For an outline of the thinking of the Conservative Government see *Investment Incentives,* Oct.1970, Cmnd.4516, HMSO, London, 1970.
81. See *H.C. Debates,* vol.820, written answers, cols. 441-2, 8 July 1971.
82. Mineral Exploration and Investment Grants Act 1972, c.9, s.1 (2). In fact no grants have so far been made for work on the continental shelf. See the *Mineral Exploration and Investment Grants Act 1971, Annual Report by the Department of Trade and Industry for the period 8th July 1971 to 31st March 1973,* pursuant to s.1 (7) of the Act, HMSO, London, 1973, para.12; id., Second Report, for the period 1 Apr.1973 to 31 Mar.1974, HMSO, London, 1974.
83. Ibid., s.1 (5).
84. Industry Act 1972, c.63, s.10 (1). S.10 (3) allows this figure to be increased to £1,400,000,000 by Statutory Instrument requiring the positive assent of both Houses of Parliament. This was done by the Ships and Mobile Offshore Installations Construction Credits (Increase of Limit) Order 1975, S.I.1975, No.138, as by 31 Mar.1974 some £642 million (including guarantees under the Shipbuilding Industry Acts) was guaranteed under this section. See *Annual Report to Parliament by the Secretary of State for the year ended 31st March, 1974,* HMSO, London, 1974, para.56. The limit itself has now been raised to £1,800 million by s.23 of the Industry Act 1975, c.68. That Act also enables guarantees to be renewed if the liability to which it relates is transferred to another company in a group (see s.24) and enables grants to be made to creditors (see s.25).

85. *The Times*, 14 Oct. 1972, p.21.
86. An idea of the way it works can be gained from the *Select Committee on Science and Technology (Sea-Bed Engineering Sub-Committee)*, Minutes of Evidence, Session 1974, H.C. Paper No.107: ii, col.783 *et seq.*, 14 May 1974.
87. See IMEG Report, p.6 and ch.3; 'Offshore Opportunities', *Petroleum Press Service*, Feb.1973, p.81.
88. See William Hall, 'N.S. Oil: What British Banks are Doing', *The Banker*, vol.123, Feb.1973, p.153. See too James McGill, 'N.S. Oil U.K. Firms Catching up in the "Spin-off" Race', *Industrial Management*, Dec./Jan.1973, p.34.
89. See *H.C. Debates*, vol.866, cols.1811-33, 21 Dec.1973.
90. Note the Secretary of State (New Departments) Order 1974, S.I. 1974, No.692. For a detailed discussion of the problems of involvement in the North Sea see *Report from the Select Committee on Science and Technology: Offshore Engineering*, Session 1974, H.C. Paper 313 of 1974, HMSO, London, 1974.
91. See 'Britain's Lost Bonanza', *Sunday Times*, 21 Apr.1974.
92. See the *Guardian*, 26 Apr.1974.
93. *The Times*, 7 Nov.1975, p.17.
94. See Peter Hillmore, 'Problem of 80 Platforms', the *Guardian*, 22 Nov.1974; *H.C. Debates*, vol.883, written answers, col.639, 20 Dec.1974. See too *Development of the oil and gas resources of the U.K.*, 1975, Appendix 13, and pp.4 and 5.
95. Department of Energy, *United Kingdom Offshore Oil and Gas Policy*, 11 July 1974, Cmnd.5696, HMSO, London, 1974.
96. Note too s.43 of the Finance (No.2) Act 1975.
97. See *Petroleum Economist*, Jan.1975, p.4.
98. See *H.C. Debates*, vol.884, cols.448 *et seq.*, 15 Jan.1975. See too below pp.29-30.
99. E.g. Amoco (see the *Guardian*, 8 Jan. 1975); Occidental (see *The Times*, 20 Jan.1975, p.17); Union Oil of California (see *The Times*, 4 Feb.1975, p.17); Conoco (see the *Guardian*, 6 Feb.1975).
100. *Petroleum Times*, 7 Mar.1975, p.3.
101. See *The Times*, 27 Feb.1975, p.17.
102. See *Petroleum Economist*, Dec.1974, p.446.
103. See P.R. Odell, 'The Future of Oil: A Rejoinder', 139 *Geographical Journal*, 436. (Note the comment by R.D.M. Simpson at 455.) Cf. Commission of the European Communities, *The Community Oil Sector medium term forecast and guidelines*, p.44. P.R. Odell and K.E. Rosing, 'Estimating World Oil Discoveries up to 1999 – the question of method', *Petroleum Times*, 7 Feb. 1975, p.26.
104. Oil Taxation Act 1975, c.22, s.1.
105. Ibid., s.2.
106. Ibid., s.10.
107. Ibid., s.8.
108. Ibid., s.9
109. *Petroleum Intelligence Weekly*, vol.14, no.24, 7 July 1975, p.7. See too *The Oil and Gas Journal*, vol.73, no.26, 30 June 1975, pp.73-8.
110. Ibid., p.8.
111. National Coal Board (Additional Powers) Act 1966, s.1 (1) (a).
112. Gas Act 1965, s.1 (2) gave the Gas Council power –

> (a) to manufacture gas, to get and acquire gas from Great Britain or elsewhere. . .

(b) to carry on such activities as it may appear to the Council to be requisite, advantageous, or convenient for them to carry on for or in connection with the exercise of their powers under. . .the foregoing paragraphs.

113. James Poole, *Sunday Times,* 14 May 1972, p.64. See too *Sunday Times,* 10 Dec.1972, p.64, on the 'Lost £3,000m.'; and the *First Report from the Committee of Public Accounts, Session 1972-3, North Sea Oil & Gas,* (together with the Proceedings of the Committee, Parts of the Minutes of Evidence, Parts of the Minutes of Evidence of Session 1971-2, and Appendices thereto), paras.26, 54-66, 87-92 and 98, HMSO, London, 1973.
114. See *H.C. Debates,* vol.787, cols.1734-41, 23 July 1969. A strange proposal as the centre of activity was then definitely on the *East.*
115. Gas Act 1972, c.60, s.2.
116. See the Petroleum and Submarine Pipe-lines Act 1975 (Commencement) Order 1975, S.I.1975, No.2079.
117. See *The Times,* 14 Dec.1975, p.1, and *The Times,* 19 Dec.1975, p.11.
118. Petroleum and Submarine Pipe-lines Act 1975, c.74, s.2 (1) (a). But see s.2 (4) (a).
119. Ibid., s.2 (1) (e). But see s.2 (4) (c).
120. Some £50,000. The BNOC also paid a further £90 million to meet outstanding loan obligations. See *The Times,* 6 Jan.1976, p.13.
121. See *Development of the oil and gas resources of the U.K.,* 1975, Appendices 9-12.
122. Petroleum and Submarine Pipe-lines Act 1975, s.9 (1).
123. *The Times,* 24 Jan.1975, p.1. The Government already holds 48.2 per cent of the equity. The Bank has in fact undertaken not to use its voting power.
124. See *The Times,* 4 Dec.1975, pp.8 and 17.
125. *The Times,* 13 Jan.1976, p.19. See too *The Oil and Gas Journal,* vol.74, no.3, 19 Jan.1976, p.22.
126. See *The Times,* 27 Feb.1976, p.17.
127. See *The Times,* 20 Feb.1976, p.17.
128. See *The Times,* 26 Nov.1975, p.23.

CHAPTER 2

1. See the *Report of the First United Nations Conference on the Law of the Sea,* Geneva, 1958, Cmnd.584, Nov.1958.
2. See *Europe: Basic Laws and Concession Contracts,* Petroleum Legislation series, Supp.22, N.Y.
3. These are the words of the Preamble. They reflect Articles 24 and 26 of the convention on the High Seas.
4. See D.W. Bowett, [1971B] 30(1) C.L.J. 50.
5. See *The Oil and Gas Journal,* vol.72, no.41, Newsletter, 14 Oct.1974.
6. See *The Third U.N. Conference on the Law of the Sea: Official Records,* 2 vols., 1st Session, New York, 3-15 Dec.1973; 2nd Session, Caracas, 20 June - 29 Aug.1974. For British proposals see *The International Sea Bed Regime: U.K. Proposals for a Convention,* The Foreign Office, London, 1972 (Cyclostyled). See too Elizabeth Young, Brian Johnson (eds.), *The Law of the Sea,* Fabian Research Pamphlet No.313, London, 1973.
7. Cf. C.F. Amerasingh, 'Basic Principles Relating to the International Regime of the Oceans at the Caracas Session of the U.N. Law of the Sea Conference', (1975) 6 *Journal of Maritime Law and Commerce* 215, 1975.
8. See the Department of Trade and Industry Map of the Fourth Round of

Licence Allocations, Mar.1972. For treaties see *Norway No.1* (1965) dated 10 Mar.1965, Cmnd.2626; *Netherlands No.1* (1965) dated 6 Oct.1965, Cmnd.2831; *Denmark No.1* (1966) dated 3 Mar.1966, Cmnd.2973. See too Richard Young, 'Offshore Claims and Problems in the North Sea', (1965) 59 *American Journal of International Law* 505, 1965; Joseph W. Morris, 'North Sea Continental Shelf: Oil and Gas Problems', (1968) 2 *International Lawyer* 191, 1968.

9. The agreement can be found at (1971) 10 *International Legal Materials* 600, 1971, and the report of the Court at [1969] *I.C.J. Reports 3.* New agreements were then reached with the UK on the median line. See *Denmark No.1* (1971), Cmnd.4882; *Federal Republic of Germany No.1* (1971), Cmnd.4881; and *Netherlands (Protocol)* (1971), Cmnd.4875, all dated 25 Nov.1971.
10. See *Norway No.1* (1965), Cmnd.2826, Article 4; *Netherlands No.2* (1965), dated 6 Oct.1975.
11. Other relevant disputes threatening are over the Falkland Islands, and, in particular, Shag Rocks (see David Fairhall, 'Rocking the Oil Boat', the *Guardian,* 6 Dec.1974) and over the Spitzbergen group of islands.
12. See Gillian Linscott, 'Green Seas' the *Guardian,* 20 July 1974, p.11. The island was annexed to the Crown by proclamation in 1955 and incorporated into Scotland by the Island of Rockall Act 1972.
13. See *The Times,* 11 July 1975, p.1.
14. See, for example, the *Guardian,* 19 July 1974, p.1; the *Daily Mail,* 19 July 1974, p.1.
15. See *The Times,* 28 Feb.1975, p.17.
16. Cf. Article 5 (6) of the Convention of the Continental Shelf, which provides: 'Neither the installations nor devices, nor the safety zones around them, may be established where interference may be caused to the use of recognised sea lanes essential to international navigation.' See to Charles Cook, 'No Entente on Channel Oil', the *Guardian,* 27 Mar.1974.
17. Section 1 of the Continental Shelf Act 1964 vests any rights exercisable by the United Kingdom outside territorial waters with respect to the sea bed and subsoil and their natural resources, except for coal, in Her Majesty.
18. Article 5 (3) provides that these zones may extend to a distance of 500 metres around such installations measured from each point of their outer edge. See too s.2 (1) and (3) of the Continental Shelf Act 1964 and the Continental Shelf (Protection of Installations) Orders, (No.1) S.I.1967, No.655; (No.2) S.I.1968 No.323; (No.3) S.I.1969 No.195; (No.4) S.I.1969 No.1322; (No.5) S.I. 1972 No.883; (No.6) S.I. 1973 No.284; as well as the similarly entitled Order SI 1976 No.332 and (No.2) S.I. 1976 No.954. See too section 3 (1) (a) of the Continental Shelf Act, and sections 8 (2)(b) and 10 (1) (a) of the Mineral Workings (Offshore Installations) Act 1971 in relation to the application of criminal and civil law to the 500 metre zone.

19. In Britain the vessel has to be capable of use 'in navigation' (see the Merchant Shipping Act 1894, s.742) which broadly speaking means that it has to be capable of proceeding under its own power. Nevertheless there are many examples of installations that meet this criterion, platforms with propelling equipment, and the more sophisticated drill ships which are fitted out to drill either through a hole in the centre of the hull or from a derrick positioned over the side.
21. 67 Stat.29; 43 U.S.C. paras.1301-3; 1311-15.
22. 67 Stat.463; 42 U.S.C. paras.1331-43.

23. A good account of the controversy is Hubert Marshall and Betty Zisk, *The Federal-State Struggle for Offshore Oil,* Inter-University Case Program No.98, Bobbs-Merrill, Indianapolis and New York, 1966. Further accounts which place more emphasis on the legal position are Oliver L. Stone, 'Legal Aspects of Marine Oil and Gas Operations', *Fifteenth Annual Institute on Mineral Law,* ed. George W. Hardy III, Louisiana State University, Louisiana, 1966, pp.21-68, and Public Land Law Review Commission, *Study of the Outer Continental Shelf Lands of the United States,* by Nossman, Waters, Scott, Krueger and Riordan of Los Angeles, Calif., Oct.1968, 2 vols., ch.3. The latter work provides an extensive analysis of the law.
24. 43 U.S.C. para.1311 (a).
25. It is precisely this point which has continued to give trouble. See *U.S.* v. *Louisiana,* 363 U.S. 1 (1960); *U.S.* v. *Florida,* 363 U.S. 121 (1960); *U.S.* v. *Louisiana,* 389 U.S. 155 (1967); *U.S.* v. *Louisiana,* 394 U.S. 1 (1969); and *U.S.* v. *Maine,* 420 U.S. 515 (1975). The result of these cases is that the seaward boundary of the states is three geographical miles measured from the coast for all states except Texas and Florida which have boundaries in the Gulf of Mexico three leagues (just over ten miles) out. See too the Act itself, 43 U.S.C. ss.1301 (b) and (c), and 1312.
26. To exactly the same extent that it would be applicable within the state on shore. Cf. *Barber* v. *New England Fish Co.* 510 F2d, 806 (Alaska, 1973). See *Mobil Oil Corporation,* 219 N.L.R.B. No.91 (1975), cf. *U.S.* v. *Alaska,* 497 F2d. 1155 (1974).
27. 43 U.S.C. paras.1302 and 1314 (a). Note that the power given to the Federal authorities under 33 U.S.C. para.403 is generally exercised in such a way as to give authority to the Secretary of the Army to order the removal or alteration of any structure if in his opinion it causes obstruction to navigation. Cf. 43 U.S.C. para.1333 (f).
28. 43 U.S.C. para.1302.
29. 43 U.S.C. para.1333 (a) (1).
30. 43 U.S.C. para.1337 (a)
31. 43 U.S.C. para.1337 (b) (3).
32. 43 U.S.C. para.1337 (a).
33. 43 U.S.C. para.1337 (b) (4)
34. 43 U.S.C. para.1334 (a) (1)
35. See 43 U.S.C. para.1340
36. 30 C.F.R. para.250.19 (see 39 F.R. 45014, 30 Dec. 1974).
37. See 30 C.F.R. paras. 250.1; 250.2(j); 250.11-12 and O.C.S. Orders, e.g. for the Gulf of Mexico, No.1 (28 Aug. 1969) para.1; No.2 (1 Jan. 1975) paras.4 and 5. Cf. 43 U.S.C. para 1333 (e) (2).
38. 43 U.S.C. para 1333 (a) (1).
39. P.L. 93-637, para.19 (f), 88 Stat. 2146; 33 U.S.C. para.1518 (f).
40. 43 U.S.C. para.1333 (a) (2) as amended by 33 U.S.C. para.1518 (f). Similar language is used in the Act of 1974 in relation to deepwater ports, which broadens the range of regulated structures used in offshore oil and gas operations. See 33 U.S.C. para.1518 (a) and (b). The term 'deepwater port' is defined as

> . . . any fixed or floating manmade structure other than a vessel, or any group of such structures, located beyond the territorial sea and off the coast of the United States and which are used or intended for use as a port or terminal for the loading or unloading and further handling of oil for transportation to any State. . .The term includes all associated components and equipment, including pipelines, pumping stations, service platforms,

mooring buoys, and similar appurtenances to the extent they are located seaward of the high water mark. (33 U.S.C. para.1502 (10).)

41. 43 U.S.C. para.1333 (d).
42. *Rodrique* v. *Aetna Casualty and Surety Co.* 395 U.S. 352 (1969), and 395 F.2d 216 (C.A.5, 1968). See too *Pure Oil Co. v. Snipes* 293 F.2d 60 (C.A.5, 1961); *Guess v. Reed,* 290 F2d 622 (C.A.5, 1961) *Continental Oil Co. v. London Steam Ship Owners' Mutual Insurance Assn. Ltd.* 417 F.2d 1030 (C.A.5, 1969); *Bertrand* v. *Forest Corp.* 441 F.2d 809 (C.A.5, La., 1971); *Nations* v. *Morris* 483 F.2d 577 (C.A.5, La., 1973), *cert. denied* 414 U.S. 1071; *Bertrand* v. *Shell Oil Co.* 489 F.2d 293 (C.A.5, La., 1973); *Oppen* v. *Aetna Ins. Co.* 485 F.2d 252 (C.A.9, Cal., 1973); *Higginbotham v. Mobil Oil Corp.* 357 F.Supp 1164 (D.C. La., 1973), *supp.op.* 360 F. Supp. 1140 (D.C. La., 1973)
43. *Pure Oil Co.* v. *Snipes* 293 F.2d 60 (C.A.5, 1961); *Oppen* v. *Aetna Ins.Co.* 485 F.2d 252 (C.A.9, Cal., 1973). Cf. the Death on the High Seas Act of 1920, 46 U.S.C. para.761 *et seq.*
44. 29 U.S.C. para.151 *et seq.*
45. 43 U.S.C. para.1333 (d).
46. 29 U.S.C. para.213 (f).
47. 29 U.S.C. para.653 (a).
48. 29 U.S.C. para.653 (b) (1).
49. 43 U.S.C. para.1333 (e) (1).
50. 43 U.S.C. para.1333 (c) (1).
51. 29 U.S.C. para.653 (b) (4).
52. Floating drilling platforms. See further p.49 below.
53. 33 U.S.C. paras.901-50.
54. 46 U.S.C. para.688.
55. For discussions of these cases see Larry P. Boudreaux Sr., 'When is an Offshore Oilfield Worker a Seaman?', (1967) 27 *Louisiana Law Review,* pp.757-84. See too E.A. Carrere, 'An Act named Jones', (1967) 15, *Louisiana Bar. Journal,* pp.95-103. Other discussions are to be found in Ralph Slovenko (ed.), *Oil and Gas Operations: Legal Considerations in the Tidelands and on Land,* Claitor's Book Store, Baton Rouge, Louisiana, 1963.
56. Don E. Kash *et al., Energy Under the Oceans. A Technological Assessment of Outer Continental Shelf Oil and Gas Operations,* Univ. of Oklahoma Press, Norman, Oklahoma, 1973, pp.104-5. This excellent study should be read for a detailed consideration of all aspects of operations on the US outer continental shelf.
57. Law of 21 April 1810 on mines, surface mines and quarries *(Bulletin des Lois,* 285) (as amended), supplemented by the Mining Law of 1903 (State Journal 1904, No.73) (as amended).
58. The Continental Shelf Mining Law of 1965 (Mignwet Continentaal Plat) (Statute Book, 1965, 428) which came into force on 1 March 1967 (see Decree of 7 Feb.1967 made under s.43).
59. See the Continental Shelf Mining Law 1965, s.1 (2), and the Decree of 7 Feb.1967 made under that section.
60. The Minerals Exploration Law of 3 May 1967 (Statute Book, 1967, 258).
61. Continental Shelf Mining Law 1965, s.12. See the Royal Decrees of 27 Jan. 1967 and 7 Feb.1967.
62. Ibid., s.26 (1) (b) and (c).
63. Ibid., ss.27 and 28.
64. Ibid., part 5.
65. Ibid., s.26 (3).
66. Decree of 13 Mar.1967 for the execution of section 26 of the Continental

Shelf Mining Act (the Continental Shelf Mining Regulations). Amended in 1968 and 1971.

67. Continental Shelf Mining Regulations, ss.9 (1) and (2), and 11.
68. Ibid., s.15.
69. Ibid., s.17.
70. Ibid., s.18.
71. Ibid., s.19.
72. Ibid., s.32 (1).
73. Ibid., ss.36-40.
74. Ibid., ss.41 and 44.
75. Ibid., 47.
76. Ibid., ss.83-91.
77. Ibid., s.97.
78. Ibid., s.98.
79. Ibid., s.102 (2).
80. Ibid., ss.111-22.
81. Ibid., ss.143-84.
82. Ibid., ss.185-91.
83. Ibid., ss.193-212.
84. Law of 3 Dec. 1964 for North Sea Installations, Art.2.
85. Ibid., Arts. 3-5.
86. See the Provisional Act relating to the Exploration for and Exploitation of Submarine Natural Resources, 31 May 1963; the Temporary Royal Decree of 15 May 1964; and the Royal Decree of 26 Feb.1965. The basic legislation giving authority over the continental shelf remains Act No.12 of 21 June 1963 relating to Exploration for and Exploitation of Submarine Natural Resources.
87. See the Royal Decree of 9 Apr.1965, s.1.
88. The Royal Decree of 8 Dec.1972, relating to Exploration for and Exploitation of Petroleum in the Seabed and Substrata of the Norwegian Continental Shelf.
89. Ibid., s.38 (e).
90. Ibid., s.57.
91. Ibid., s.47.
92. Ibid., s.53.
93. Ibid., s.51.
94. The Royal Decree of 25 Aug.1967 relating to safe practice etc. in exploration and drilling for submarine petroleum resources. See for example ss.4 and 15.
95. Ibid., s.7.
96. Ibid., s.17.
97. Ibid., s.18.
98. Ibid., ss.26-31.
99. Ibid., s.23.
100. Ibid., s.41.
101. Ibid., s.101.
102. Ibid., s.43.
103. Ibid., s.2. See the Royal Decrees of 11 July 1969 and 1 Apr.1973.
104. Act No.259 of the 9 June 1971 concerning the Continental Shelf, as amended by Act No.278 of 7 June 1972. The Act came into force on 1 July 1971, repealing Art.3 of the Royal Decree of the 7 June 1953.
105. Act No.181 of 8 May 1950 on Exploration and Utilization of Raw Materials in the Subsoil of the Kingdom of Denmark.
106. See the Order Respecting the Monopoly of Prospecting for and Extraction of Hydrocarbons etc. in the Danish Subsoil, Order No.372 dated 7 Nov.1963. The original Order was published as Order No.270 of the Ministry of Public Works of 16 July 1962. The agreement was also subject to a Protocol of

3 Mar.1972 respecting the extent to which Danish industry should benefit from the activities of the concessionaire. After the concession was granted, Shell and Gulf Oil joined Møller in forming the Dansk Undergrunds Consortium and in 1965 Texaco and Chevron also became members, although as three different Dansk groups were formed, the companies participate differently in each of three areas.

107. Act. No.259, *supra,* Art.2, para.1, final sentence.
108. See Order 372, *supra,* s.6 (3).
109. Ibid., s.8.
110. Act No.259, *supra,* Art.4 (1).
111. Ibid., Art.4 (2).
112. Ibid., Art.3 (1).
112a. See Decree No.65-1049 of 29 Nov.1965.
113. Law No. 68-1181 of 30 Dec.1968 *(Journal Officiel,* 31 Dec.1968, p.12404), Art.6.
113a. Ibid., Arts.15-19.
114. Ibid., Arts.20-3.
115. Ibid., Arts.2 and 6, and see too Decree No.71-360 of 6 May 1971 *(Journal Officiel,* 15 May 1971, p.4692), Arts.1-15; Decree No.71-362 of 6 May 1971 *(Journal Officiel,* 15 May 1971, p.4695); and Decree No.70-988 of 29 Oct. 1970 *(Journal Officiel,* 30 Oct.1970, p.10032).
116. Law No.68-1181, *supra,* Art.3.
117. Ibid., Art.10.
118. Ibid., Art.10.
119. Ibid., Art.11.
120. Ibid., Art.11.
121. Ibid., Art.4.
122. Ibid., Art.5.
123. Decree No.71-360, *supra,* Art.23.
124. Law No.68-1181, *supra,* Art.35.
125. Decree No. 71-360, *supra,* Art.24.
126. Law No.68-1181, *supra,* Art.9.
127. Ibid., Art.36.
128. Law of 13 June 1969 on the Continental Shelf *(Moniteur Belge,* 8 Oct.1969).
129. See the Law for the Temporary Regulation of the Rights to the Continental Shelf (Gesetz zur vorläufigen Regelung der Rechte am Festlandsockel) of 24 July 1964 *(Bundesgesetzblatt* 1, 497) as amended by Art.150 (2) (16) of the Law of 24 May 1968 *(Bundesgesetzblatt* 1, 581); Art.190 of the Law of 2 Feb.1974.*(Bundesgesetzblatt* 1, 469); and the Law of 2 Sept.1974 *(Bundesgesetzblatt* 1, 2149).
130. Petroleum and Other Minerals Development Act 1960, No.7 of 1960, s.4.
131. Ibid., s.6 (1) and (2).
132. Ibid., ss.8 (2), 9 (3), 13 (2) (b), and 19 (2).
133. Ibid., ss.8 (3), 9 (4), 13 (2) (a), 19 (2) and 53.
134. See ibid. ch.6 and 7.
135. Ibid., s.52 (1) (a) (ii) and (viii).
136. Ibid., s.52 (2) (b).
137. Continental Shelf Act 1968, No.14 of 1968, s.4 (2).
138. Ibid., s.2 (1).
139. Ibid., s.2 (3).
140. Ibid., s.5 (1).
141. See ibid., ss.5 (esp. s.5 (3)) and 12.
142. Ibid., s.6.
143. Ibid., ss.10 and 11.

144. See pages 54-60 below.
145. I.e. a wrong under the Civil Liability Act 1961, No.41 of 1961. See the Continental Shelf Act 1968, s.3 (2) (c).
146. Continental Shelf Act 1968, s.3.
147. Ibid., s.13 (1).
148. Ibid., s.3 (3).
149. Continental Shelf Act 1964, c.29, s.1 (1).
150. See the Continental Shelf (Designation of Areas) Order 1964, S.I. 1964 No.697; the Continental Shelf (Designation of Additional Areas) Order 1965, S.I. 1965 No.1531, and the similarly entitled Orders providing for additional areas, S.I. 1968 No.891; S.I. 1971 No.594; S.I. 1974 No.1489, and S.I. 1976 No.1153.
151. There is still considerable dispute as to the exact limit of territorial waters which may or may not be solved by the UN Conference on the Law of the Sea. In relation to the UK licensing system it is not important, although the distinction is relevant for certain other purposes. At the 1958 Conference the UK Foreign Office declared that it would go on treating the limit as being three miles until some further agreement was reached (See Cmnd.584, p.6, para.6). This limit finds support in the definition found in s.7 of the Territorial Waters Jurisdiction Act 1878, and similar references are made in the Customs and Excise Act 1952, ss.75, 76 and 173 (3). See too the Cornish Submarine Mines Act 1858. See further *Direct United States Cable Co. v. Anglo-American Telegraph Co.* (1877) 2 App. Cas. 394; *The Fagerness* [1927] P.311, *R. v. Kent Justices, ex parte Lye* [1967] 1 All E.R. 560, and *Post Office* v. *Estuary Radio Ltd.* [1967] 3 All E.R. 663, 679 (C.A.). There is also argument that the limit should be measured not from the low-water mark, but from a base-line (see the *Anglo-Norwegian Fisheries Case* [1951] I.C.J. Reports 116 and comments thereon).
152. 24 & 25 Geo. V, c.36.
153. Petroleum is defined in s.1 (2) of the Petroleum (Production) Act 1934, which reads:

> '. . .the expression"petroleum" includes any mineral or relative hydrocarbon and natural gas existing in its natural condition in strata, but does not include coal or bituminous shales or other stratified deposits from which oil can be extracted by destructive distillation.'

154. I.e. the rights conferred by s.1 (1) of the Continental Shelf Act.
155. The functions of the Board of Trade were transferred to the Department of Trade and Industry by the Secretary of State for Trade and Industry Order 1970, S.I. 1970 No.1537 and are now transferred to a Secretary of State by the Secretary of State (New Departments) Order 1974, S.I. 1974 No.692 and are exercised by the Secretary of State for Energy.
156. Emphasis added. The general terms and conditions of contracts with the Government are discussed in Colin Turpin, *Government Contracts,* Penguin, London, 1972. Somewhat surprisingly, consideration of contracts under the Petroleum (Production) Act seems to have been omitted from this work. In the Fourth Round of Allocations by bidding for Competitive Tender Blocks, which opened on 20 Aug.1971, the Department of Trade and Industry announced the successful bids on 25 Aug. for 15½ blocks at a total price of £37,213,653.99 (an average price of £2,400,800.903 per block). In fact, the Shell/Esso partnership paid £21,050,001 for block 211/21, which later turned out not to be oil producing. At the other end of the scale Chieftain Exploration (UK) Ltd paid £8,000 for 1½ blocks off East Anglia in an already well-explored region. (See DoTI *Press Notice,* 25 Aug.1971, Ref.No.734). Of

the 421 blocks offered for discretionary allocation 271 were applied for by the closing date of 20 Aug. Over 200 firms were involved in a total of 92 applications. Altogether some 267 blocks have now been awarded on a discretionary basis. It is interesting to note that 43 per cent of the total territory involved in the fourth round was allocated to *indigenous* British companies. (See DoTI *Press Notice,* 22 Dec.1971, Ref.No.115.) The British share in previous rounds of licensing was 30 per cent in 1964, 37 per cent in 1965 and 40 per cent in 1970. (See DoTI *Press Notice,* 15 March 1972, Ref. No.241.) Altogether the Government received a total of £89,257,293 from royalties, fees and tender prices in connection with exploration and exploitation of the continental shelf between 1 Apr.1964 and 30 Sept.1974. See *North Sea Oil and Gas: A Report to Parliament,* DoTI, HMSO, London, 1973, table 5; *First Report of the Committee of Public Accounts Session 1972-3; North Sea Oil and Gas,* para.7, and the *Annual Reports to Parliament* under s.5 (1) of the Petroleum (Production) Act 1934. Details of current holdings are set out in the *Annual Reports to Parliament* under s.1 (5) of the Continental Shelf Act 1964. A fifth round of licensing is to take place in 1976, and most of the blocks on offer will be in the Celtic Sea, the Western Approaches, and west of Orkney and the Shetlands. The terms are bound to include built-in participation by the BNOC. See *The Times,* 4 Sept.1975, p.17. These areas are not thought to be too promising. See *The Oil and Gas Journal,* vol.73, no.40, 6 Oct.1975, p.38.

157. Petroleum (Production) Act 1934, s.6 (1) (d).
158. The Petroleum (Production) Regulations 1966, S.I. 1966 No.898.
159. By the Petroleum (Production) (Amendment) Regulations 1971, S.I. 1971 No.814, the Petroleum (Production) (Amendment) Regulations 1972, S.I. 1972 No. 1522, the Petroleum and Submarine Pipe-lines Act 1975, ss.17, 18 and 19 and Schedules 2 and 3, the Petroleum (Production) Regulations 1976, S.I. 1976, S.I. 1976 No.276.
160. The Petroleum (Production) Regulations 1966 (as amended), Regulation 4. Note in this connection the Directive of the Council of the EEC of 13 March 1969 (No.69/82) (see *Journal Officiel* No.L68 19 Mar.1969, p.4).
161. The Regulations of 1966 replaced, among others, the Petroleum (Production) (Continental Shelf and Territorial Sea) Regulations 1964, and were made after the 'Sea Gem' disaster (on which see later chapter) on the 21 July. They reflect the current anxiety. Schedule 2 to the Regulations, which has been slightly amended by the Petroleum (Production) (Amendment) Regulations 1971, S.I. 1971 No.814, deals, in accordance with Regulation 5, with the form of the application itself. The schedule requires the applicant to state with a great deal of thoroughness details of himself, such as usual place of residence, and evidence of nationality. If the applicant is a company, it is required to give the following details:

(a) place of incorporation
(b) principal place of business
(c) the registered office
(d) the place of central management or control
(e) particulars of the members of the board of directors or other governing body, their names, addresses and nationalities.
(f) if the application is for a production licence (described elsewhere), particulars must be given of the capital issued and authorised, that is to say, for each class of capital details of the amount authorised, the amount issued, and the voting rights of each class, must be given. Furthermore, particulars of all holdings of not less than 5 per cent in number or value of any class of capital which has been issued, including the full name of the

holder or holders, the class of the holding, the amount, and the nationality of the holder(s). Similar particulars must be given of the capital issued to the bearer.

With the articles and memorandum of the company (obtainable anyway, but which may be deposited instead of giving details of the voting rights of each class) the Department would appear to have fairly thorough knowledge of both ownership and control.

162. See the Petroleum (Production) Regulations 1966, S.I. 1966 No.898, Regulation 10. There are five different kinds of licence described in the Schedules to the regulations: production licences for both seaward and landward areas, exploration licences also for either seaward or landward areas, and methane drainage licences. Schedules 4 and 5, which relate to seaward areas, are of real relevance here.
163. See the Petroleum and Submarine Pipe-lines Act 1975, ss. 17-19, and Schedules 2 and 3.
164. See, for example, ibid., Schedule 2, Part 2, para.22.
165. See the Petroleum (Production) Regulations (as amended), Schedule 4, para.24, and Schedule 5, para.11.
166. In an exploration licence called the 'exploration area'.
167. Regulations relating to safety, health and welfare have now been published, see The Offshore Installations (Operational Safety, Health and Welfare) Regulations 1976, S.I. 1976 No.1019 and note esp. Regulation 7.
168. The Petroleum (Production) Regulations (as amended), Schedule 4, para.34, and Schedule 5, para.18.
169. Ibid., Schedule 4, para.39; Schedule 5, para.21. Revocation may occur for a variety of reasons including financial instability, breach or non-observance of any of the terms and conditions of the licence, if the licensee ceases to be a UK citizen resident in the UK, or in the case of a company, ceases to have its central management and control in the UK.
170. This situation is to be contrasted with the other well-known area in which a Minister looks after the interests of third parties under agreements between the Minister of Transport and the Motor Insurer's Bureau. The MIB/MoT agreement permits the third parties to sue under the agreement. See *An Agreement between the M.o.T. and the M.I.B. dated 21st April 1969,* HMSO, London, and the *Agreement between the Secretary of State for the Environment and the M.I.B. dated 1st Feb.1971,* HMSO, London. As to the legality of the proceedings see *Coward* v. *M.I.B.* [1963] 1 Q.B. 259 per Upjohn L.J.; *Hardy* v. *M.I.B.* [1964] 2 Q.B. 745 per Lord Denning M.R.; *Albert* v. *M.I.B.* [1971] 3 W.L.R. 291 per Viscount Dilhorne.
171. Subject to questions of a *jus quaesitum tertio* arising in Scotland. See D.M. Walker, *Principles of Scottish Private Law,* Oxford U.P., London, 1970, vol.I, pp.567-9; T.B. Smith, *A Short Commentary on the Law of Scotland,* W. Green, Edinburgh, 1962, pp.777-84. It is hoped that the licences issued under the Fifth Round of Allocations in 1976 will improve the position of third parties.
172. The Act did not come into force immediately, see the Mineral Workings (Offshore Installations) Act 1971, c.61, s.14 (2). All the sections apart from 4 and 5 came into force on 1 May 1972, and sections 4 and 5 came into force on 1 Aug.1972. See the Mineral Workings (Offshore Installations) Act (Commencement) Order 1972, S.I. 1972 No.644 (C.11), dated 21 Apr.1972.
173. Even those oil men with wide experience of offshore work were really very shocked by conditions in the North Sea. See, for example, M.J. Wells, 'Hard

Lessons Learned in the North Sea', *World Petroleum,* Mar.1966, p.23.
174. Eric Ogden, MP for Liverpool, West Derby, *H.C. Debates,* vol.816, col.653, 28 Apr.1971.
175. *Report of the Inquiry into the causes of the Accident to the Drilling Rig Sea Gem* (President: J. Rowland Adams, QC), Oct. 1967, Cmnd. 3409 (referred to henceforward as the *'Sea Gem' Report),* para.10.2 (i).
176. See *H.C. Debates,* vol.816, col.645 *et seq.,* 14 July 1971.
177. More specifically s.1 (2) (a) reads: 'Waters to which this Act applies' means the waters in or adjacent to the United Kingdom up to the seaward limits of territorial waters, and the waters in any designated area within the meaning of the Continental Shelf Act 1964. By s.1 (2) (b): 'inland waters' means waters within the United Kingdom, other than estuaries and tidal rivers.
178. *Quaere* as to whether there is not a 'gap' here in so far as the Continental Shelf Act vests rights in the sea bed that includes some non-minerals. See the Continental Shelf Act 1964, s.1 (1).
179. Mineral Workings (Offshore Installations) Act 1971, s.1 (3) that part which excludes an installation from the Act if it is 'connected with dry land by a permanent structure providing access at all times and for all purposes' was inserted in order that there should not be confusion about structures like the Brighton pier. (This explanation was provided by a senior civil servant in charge of the legislation.) The definition also makes a distinction between offshore oil-rigs and pirate radio 'structures': see s.2 (1) of the Marine etc. Broadcasting Act 1967.
180. Ibid., s.1 (3) (b). Note too s.12 (3), as amended by s.44 (3) of the Petroleum and Submarine Pipe-lines Act 1975.
181. See the Mineral Workings (Offshore Installations) Act 1971, s.7 (7) (a) and s.10 (2) (a).
182. See *H.L. Debates,* vol.315, col.1267, 2 Mar.1971.
183. But see the Mineral Workings (Offshore Installations) Act 1971, s.7 (7) (b).
184. The Offshore Installations (Registration) Regulations 1972, S.I. 1972 No.702 draw a distinction in para.2 (1) between a 'mobile installation', i.e. one that can be moved, and a 'fixed installation'. See too s.12 (1) of the Industry Act 1972.
185. See too the Petroleum and Submarine Pipe-lines Act 1975, s.44 (1) (b).
186. Regulations made under s.44 (4) of the Petroleum and Submarine Pipe-lines Act 1975 may make provision as to what things are capable or incapable of being manned by one or more persons.
187. Petroleum and Submarine Pipe-lines Act 1975, Part 3. The definition of 'controlled waters' is given in s.20 (2), see s.48 (1).
188. Ibid., ss.20, 21, 24 and 25. For the application see s.21 (1) and Schedule 4, esp. para.1.
189. Ibid., s.22.
190. Ibid., s.23.
191. Ibid., s.21 (3).
192. Ibid., s.26 (1). See too s.46.
193. See pages 144-5 below.
194. Petroleum and Submarine Pipe-lines Act 1975, s.33 (1).
195. Ibid., s.33 (1).
196. Ibid., s.33 (2).
197. Ibid., s.31 (3).
198. Ibid., s.31 (1).
199. Ibid., s.31 (2).
200. Ibid., s.26 (2).
201. Ibid., s.26 (3).

202. Nevertheless, the Act seems to have been somewhat surprisingly overlooked by even the specialist works on statutes. See P. St.J. Langan, *Maxwell on the Interpretation of Statutes,* 12th ed., Sweet & Maxwell, London, 1969, esp. pp.169-75; S.G.C. Edgar, *Craies on Statute Law,* 7th ed., Sweet & Maxwell, London, 1971, esp. ch.18; Gerard Dworkin, Odgers' *Construction of Deeds and Statutes,* 5th ed., Sweet & Maxwell, London, 1967, see pp.407-13. The Mineral Workings (Offshore Installations) Act was, of course, passed after all these works were written.
203. See *Maxwell,* 12th ed., p.171. Approval for this statement, taken from an earlier edition, is to be found in *Tomalin* v. *S. Pearson & Son Ltd.* [1909] 2 K.B. 61 (C.A.) per Cozens-Hardy M.R. at p.64; and in *Krzus* v. *Crow's Nest Pass Coal Co. Ltd.* [1912] A.C. 590 (P.C.) per Lord Atkinson (delivering the judgement of the Court) at pp.596-7. For further examples of the general rule see: *Schwartz* v. *India Rubber, Gutta Percha & Telegraph Works Co. Ltd.* [1942] 2 K.B. 299 (C.A.); *Attn.-Gen. for Alberta* v. *Huggard Assets Ltd.* [1953] A.C. 420 per Lord Asquith at p.441; *C.E.B. Draper & Son Ltd.* v. *Edward Turner & Son Ltd.* [1964] 3 All E.R. 148, per Lord Denning M.R. at p.150. See too *R.* v. *Jameson* [1896] 2 Q.B. 425; *Yorke* v. *British and Continental Steamship Co. Ltd.* (1945) 78 Ll.L.R. 181. The problem of the limits of territorial waters has already been referred to.
204. E.g. the Royal Marriages Act 1772 (see the *Sussex Peerage Case* (1844) 11 Cl. & F. 85); the Marriage Act 1835 (see *Brook* v. *Brook* (1861) 9 H.L.C. 193 per Lord Campbell L.C. at p.210; *Re De Wilson* [1900] 2 Ch. 481; *Pugh* v. *Pugh* [1951] P. 482); the Personal Injuries (Emergency Provisions) Act 1939 (see *Howgate* v. *Bagnall* [1951] 1 K.B. 265). Plainly, most of these are special cases.
205. *Tomalin* v. *S. Pearson & Son Ltd.* [1909] 2 K.B. 61; *Yorke* v. *British and Continental Steamship Co. Ltd.* (1945) 78 Ll.L.R. 181.
206. [1963] A.C. 48 at p.67.
207. Offences against the Person Act 1861, s.9. See *R.* v. *Page* [1954] 1 Q.B. 170. See too the Perjury Act 1911, s.8. Others include bigamy (see s.57 of the Offences against the Person Act 1861, and *R.* v. *Earl Russell* [1901] A.C. 446) and treason (see *R.* v. *Casement* [1917] 1 K.B. 98). See too the Slave Trade Act 1824; and the Army Act 1955, s.70; the Air Force Act 1955, s.70; and the Naval Discipline Act 1957, s.42. For a recent case on blackmail see *Treacy* v. *D.P.P.* [1971] 1 All E.R. 110 (H.L.). For a full discussion of the subject see Glanville Williams, 'Venue and the Ambit of Criminal Law' [1965] 81 L.Q.R. 276, 395 and 518.
208. Rules of the Supreme Court, O.11, r.1. See too Lawrence Collins, 'Some Aspects of Service out of the Jurisdiction' [1972] 21 I.C.L.Q. 656.
209. The grant is in all cases discretionary.
210. Rules of the Supreme Court, O.11, r.1 (1) (c).
211. Which has been approved in a working paper of the Law Commission. See 'Territorial and Extra Territorial Extent of the Criminal Law', *Law Commission Working Paper* No.29, 12 May 1970, paras.18 and 36.
212. In his article 'Venue and the Ambit of Criminal Law' [1965] 81 L.Q.R. 276, 395 and 518, Professor Glanville Williams refers (at p.417) to the possibility of conflict between the various systems of law in the United Kingdom. This would appear to have been resolved by the Continental Shelf (Jurisdiction) Order 1968, S.I. 1968 No.892, para.2.
213. S.8 (1) and (2) of the Mineral Workings (Offshore Installations) Act 1971 extends the application of both parts of s.3 of the 1964 Act to installations in territorial waters. Presumably it was intended to clarify the position.
214. The Continental Shelf (Jurisdiction) Order 1968, S.I. 1968 No.892, which

replaces the similarly titled Order of 1965. See too the Continental Shelf (Jurisdiction) (Amendment) Order 1971, S.I. 1971 No.721, the Continental Shelf (Jurisdiction) Amendment) Order 1974, S.I. 1974 No.1490, and the Continental Shelf (Jurisdiction) (Amendment) Order 1975, S.I. 1975 No.1708.

215. Ibid., para.2. The Order divides the UK Continental shelf into three areas in which English, Scottish and Northern Ireland laws are to apply. Similar wording is applied for the other areas.

216. Ibid., para.3. Again, similar provision is made for the Scottish and Northern Ireland areas.

217. Though not necessarily with success, see *C.E.B. Draper & Son Ltd.* v. *Edward Turner & Son Ltd.* [1964] 3 All E.R. 148.

218. The phrase 'acts or omissions' was considered in *Squire* v. *Bayer & Co.* [1901] 2 K.B. 299 in connection with s.1 (1) of the Truck Act 1896. In that case a rule that provided that 'all workers shall observe good order and decorum while in the factory' was held to comply with the terms of the Act under which no deductions from wages could be made unless 'the contract specified the acts or omissions in respect of which the fine may be imposed'. The court decided that although the rule did not specify the 'acts or omissions', if it was interpreted with reference to the requirements of the particular business it nevertheless complied with the Act. The decision can hardly be called a strong precedent.

219. Crown Proceedings Act 1947, s.10 (1).

220. 6 Edw. VII, c.47, s.5 (3). Repealed by the Industrial Relations Act 1971, s.169 and Schedule 9.

221. I.e. an analogy with the situation where the worker is deemed to be outside 'the course of his employment'. See, for example, *Uddin* v. *Associated Portland Cement* [1965] 2 Q.B. 582; *Hilton* v. *Thomas Burton* [1961] 1 W.L.R. 705. See too *Davidson* v. *Handley Page Ltd.* [1945] 1 All E.R. 235, per Lord Greene at p.237. Presumably this test would apply anyway in an action for damages against the employer.

222. Note that provision is made by s.11 (1) and (2) of the Mineral Workings (Offshore Installations) Act 1971, and s.30 (1) of the Petroleum and Submarine Pipe-lines Act 1975, for actions for any breach of statutory duty imposed by or under the 1971 Act, or by regulations made under the 1975 Act, causing personal injury, and for the application of the Fatal Accidents Act 1976 in the event of death.

223. See the Factories Act 1961, s.175. There is, however, evidence that the Factories Act has been applied to offshore installations. (Private correspondence between the Department of Employment and the legal department of British Petroleum.) It is difficult to see upon what grounds. In *Cox v. S. Cutler & Sons* [1948] 2 All E.R. 665, coal gas was held to be an 'article' within the equivalent section 151 (1) of the 1937 Act, but whereas coal gas is 'made' from coal, natural gas and oil would seem on this analogy to fall outside either 'making of any article' (s.175 (1) (a)) or 'adapting for sale' (s.175 (1) (c)), as would be the case on shore.

224. Occupiers' Liability Act 1957, s.1 (3). This is further discussed below. See too s.2 (1), s.3 (1) (c) and (d) of the Marine etc. Broadcasting (Offences) Act 1967 for an example of offshore 'structures'.

225. Cf. s.30 (2) of the Petroleum and Submarine Pipe-lines Act 1975.

226. See *Maxwell,* pp.177-83; *Odgers,* pp.407-13, esp. 411-12.

227. Under the Copyright Act 1842, see *Routledge* v. *Low* (1868) L.R. 3 H.L. 100; and *Falcon* v. *Famous Players Film Co.* [1926] 2 K.B. 474 (C.A.) Under the Fatal Accidents Acts, see *Davidsson* v. *Hill* [1901] 2 K.B. 606.

228. Under the Copyright Act 1709, see *Jeffreys* v. *Boosey* (1854) 4 H.L.C. 815.

229. Except for questions such as the payment of income tax.
230. *In re Sawers, ex parte Blain* (1879) 12 Ch.D. 522 at p.526. (Emphasis added.) Approved in *Cooke* v. *Charles A. Vogeler Co.* [1901] A.C. 102 (H.L.), and in *In re Dulles Settlement, Dulles* v. *Vidler* [1951] Ch. 265.
231. See too s.12 (4) of the Mineral Workings (Offshore Installations) Act 1971 which repeats this subsection but with a substitution of the words 'this Act' for the words 'the said section 3'. Cf. s.33 (9) of the Petroleum and Submarine Pipe-lines Act 1975.
232. See *Maxwell*, pp.224-5.
233. See the Employment Protection Act 1975, c.71, s.127.
234. Information supplied by the legal department of Shell International.
235. See the Industrial Relations Act 1971, ss.5 (1) and 22 (1).
236. There has been no specific test of the meaning of this word in the courts. It has in fact fallen to be decided under s.1 (1) of the Maintenance Orders Act, 1950. See, *inter alia*, *Macrae* v. *Macrae* [1949] 2 All E.R. 34, and *Lewis* v. *Lewis* [1956] 1 All E.R. 375. In the latter case Willmer J. said, at p.378, 'I think myself it is a matter to which has to be viewed in the context of all the other circumstances of the case'. The Department of Employment took the view that s.27 (2) operated to exclude oil-rig workers. (Private correspondence between the Department of Employment and the legal department of British Petroleum.)
237. Cf. *C.E.B. Draper & Son Ltd.* v. *Edward Turner & Son Ltd.* [1964] 3 All E.R. 148, per Lord Denning M.R. at p.150.
238. See *H.C. Debates*, Standing Committee E, *Official Report*, Session 1974 (Trade Union and Labour Relations Bill), col.716, 25 June 1974.
239. Employment Protection Act 1975, c.71, s.127 (3) (a).
240. Ibid., s.127 (3) (d).
241. Ibid., s.127 (3) (b).
242. Ibid., s.127 (3) (e) and (f).
243. See the Health and Safety at Work etc. Act 1974, s.84 (3). Discussed at pp.145-6 below.
244. See the Sex Discrimination Act 1975, s.10 (5) and (6). Discussed at pp.82-4 below.
245. The Employment Protection (Offshore Employment) Order 1976 S.I. 1976 No.766.

CHAPTER 3

1. *The Sociology of Industrial Relations*, Longman, London, 1971, p.19.
2. Some companies such as Shell give all their employees 'staff' status to reduce conflict in a closed situation. The basic distinction nevertheless remains.
3. Seven days on, seven days off is standard in the Norwegian sector; has become more common on oil-rigs off Aberdeen. British Petroleum experimented on the 'Sea Quest' with a three-shift system of eight hours, but this was abandoned, as was 14/7.
4. Accommodation is discussed in a later chapter. It must be mentioned, however, that some rigs work on the 'hot bed' principle: two men on different twelve-hour shifts use the same bed.
5. Mineral Workings (Offshore Installations) Act 1971, s.4.
6. See the *'Sea Gem' Report*, paras.8.4 and 10.2.
7. Many drillers in fact have engineering degrees.
8. It takes about five, sometimes seven, years to become proficient, according

to ODECO and Conoco.
9. Reading & Bates (Offshore) Ltd, a company that owns, operates and builds offshore oil-rigs.
10. See the *I.M.E.G. Report,* pp.32-33.
11. See *Development of the oil and gas resources of the U.K.*, 1975, Appendix 14.
12. Ibid., Appendix 14, note 3.
13. See the *I.M.E.G. Report,* p.64, which put wastage as high as 50 per cent in some cases, and the Petroleum Industry Training Board, *Annual Report 1972/3,* p.30. But see *The Press and Journal,* Aberdeen, 29 Aug.1972, p.14.
14. See J.M.M. Hill, *The Seafaring Career,* Tavistock Institute, London, 1972; and the Report presented to the Rochdale Committee of Inquiry into the Shipping Industry by Social Surveys (Gallup Polls) Ltd, *Attitudes of Seafarers to their Employment,* 3 vols., Board of Trade, London, 1970.
15. In the case of Reading & Bates 85 per cent (company estimate). This figure will probably be exceeded at British Petroleum, and in the case of Shell.
16. See Hill, op.cit. The same conclusion is reached by P.G. Hollowell in *The Lorry Drivers,* Routledge & Kegan Paul, London, 1968.
17. Table 8 of vol.3 of *Attitudes of Seafarers to their Employment,* shows that 66 per cent of former ratings gave variations on this phrase for joining the industry.
18. See Hill, op.cit., paras.83 and 84.
19. See the Industrial Training Act 1964, s.1 and the Industrial Training (Petroleum Board) Order 1967, S.I. 1967 No.648 (Schedule 1 replaced by S.I. 1970 No.205). See now the Employment and Training Act 1973, esp. s.6 and Schedule 2. Under s.127 (1) of the Employment Protection Act 1975, and the Employment Protection (Offshore Employment) Order 1976, S.I. 1976 No.766, para.3, the Industrial Training Act 1964 is extended in relation to employment offshore in either territorial waters or in designated areas in connection with the sea bed or subsoil or the exploitation of their natural resources. The main difference is that contributions can now be collected from those companies which employ people only on work outside territorial waters, as distinct from those companies whose activities were already covered by the Petroleum Industry Training Board because they also employ people to work 'in Great Britain'. In effect, however, it merely endorses the present approach of the PITB. (See S.I. 1970 No.205, Schedule 1, para.1.)
20. See Petroleum Industry Training Board, *News 11,* Mar.1972, p.4.
21. PITB, *News 14,* Nov.1973, p.3.
22. See PITB, *Annual Report,* 1973/4.
23. See PITB, *News 13,* Dec.1972, p.2.
24. *IMEG Report,* ch.8. A Special Report on Education and Training by Professor L.J. Rydell, of University College, London.
25. *IMEG Report,* p.14.
26. *Education and Training for Offshore Development,* Report of an Inter-Departmental Working Party, Department of Employment, 1973.
27. J. Fyfe and D. Begg, *Deep Diving in the United Kingdom,* Training Services Agency, 1974. Cf. Manpower Services Commission (Chairman: Sir Denis Barnes), *Training Services Agency: A Five Year Plan,* HMSO, London, 1974, p.16.
28. See Petroleum Industry Training Board, *News 13,* p.2. See too The Petroleum Training Association – North Sea, *Offshore Crane Driver Training.*
29. See PITB, *Annual Reports.*
30. See PITB, *News 15,* Feb.1975, p.2.

31. For full discussion of the Act (although not in application to the continental shelf) see B.A. Hepple, *Race, jobs and the law in Britain,* 2nd ed., Penguin, London, 1970; and Anthony Lester and Geoffrey Bindman, *Race and Law,* Penguin, London, 1972.
32. Meaning to treat less favourably than he would treat other workers, another person on grounds of colour, race, or national origins. See the Race Relations Act 1968, s.1.
33. Race Relations Act 1968, s.3 (1) (a).
34. Ibid., s.3 (1) (b).
35. I.e. sections 2 and 3 of the Act.
36. See the Race Relations Act 1968, s.19, and note subsection (b).
37. Sex Discrimination Act 1975, s.10 (1).
38. See ibid., ss.62 and 63.
39. Note that s.14 of the Act at present makes it unlawful for an industrial training board to discriminate against a woman in giving her access to training courses. This would appear to apply to those courses offered by the PITB relevant to offshore employment even though no Order has as yet been made.
40. See the Sex Discrimination Act 1975, s.8 (6) and Schedule 1, Part 1, paras.2 and 4, and the Sex Discrimination Act 1975 (Commencement) Order 1975, S.I. 1975 No.1845.
41. Disabled Persons (Employment) Act 1944, s.21 (1).
42. Ibid., s.22 (2).
43. *H.C. Debates,* vol.816, col.660 (Kevin McNamara), 28 Apr. 1971.
44. Part 8 of the *Institute of Petroleum Model Code of Safe Practice in the Petroleum Industry,* Institute of Petroleum, London, 1964. Referred to hereafter as 'the Code'.
45. Under instruction No.1.
46. Para.1.1 of *the Code.*
47. See para.8.4 of the *'Sea Gem' Report.*
48. Ibid., para.8.4.
49. Ibid., para.8.3.
50. Ibid., para.10 (2) (iii). The Tribunal also recommended a public address system so that the person in command could give instructions to everyone, and the keeping of a log book to record all events happening on board relevant to safety.
51. Ibid., para.8.4.
52. Ibid., para.10 (2) (ii).
53. Which relates only to Production Licences. In fact nearly all licences are Production Licences which give the holder the right to 'search and bore for, and get, petroleum. . .', as distinct from the exploration licence which only gives the holder the right to 'search for petroleum. . .which. . .shall not include. . .any right to drill wells for production. . .or any other well of a depth exceeding 350 metres. . .'. See the Petroleum (Production) Regulations 1966, S.I. 1966 No.898.
54. The background to the provisions is unwittingly revealed by the reference to the manager in the side-notes and index as a 'Master'. In fact the term was only changed in the text during the Report stage of the passage of the Bill through the House of Commons (see *H.C. Debates,* vol.821, col.667 *et seq.,* 14 July 1971). There had been criticism of the analogy with Merchant Shipping legislation (see *H.C. Debates,* vol.816, col.660 (Kevin McNamara) and 670 (John Prescott), 28 Apr.1971), and the change was made 'to get away from the fact that it was a piece of merchant navy legislation and to make it clear that it was a piece of industrial legislation'. (Nicholas Ridley,

Under-Secretary for Trade and Industry, *H.C. Debates*, vol.821, col.667, 14 July 1971.)

55. See para.10.2 (iii) of the *'Sea Gem' Report.*
56. See Figures 1-4 above.
57. The Offshore Installations (Managers) Regulations 1972, S.I. 1972 No.703, which came into force on 31 Aug.1972. They also exclude the application of ss.4 and 5 of the Act (under s.4 (7) and s.5 (9)) to installations registered as vessels which are dredging installations or which are in transit from a station, or in relation to installations which are unmanned (Regulation 5). This confirms the view expressed earlier that the Mineral Workings (Offshore Installations) Act 1971 is only applied to such vessels whilst actually engaged in drilling.
58. Ibid., Regulation 2 (1) and Schedule 1.
59. Ibid., Regulations 3 (1) and (2).
60. The enforcement subsection for the whole section.
61. See *H.C. Debates*, vol.816, col.660 (Kevin McNamara), 28 Apr.1971.
62. *H.C. Debates*, vol.816, col.676 (Nicholas Ridley), 25 Apr.1971.
63. Para.10.2 (vi) of the *'Sea Gem' Report.*
64. The Offshore Installations (Operational Safety, Health and Welfare) Regulations 1976, S.I. 1976 No.1019, Regulations 19, 21, 30 and 31.

CHAPTER 4

1. A good account of the way the oil industry has developed, and the way in which new companies have become involved, is Neil H. Jacoby, *Multinational Oil*, Macmillan, New York, 1974, esp. ch.7.
2. These details can be found in Appendix 9 to *Development of the oil and gas resources of the U.K.*, 1975, op.cit.
3. See *The Times*, 18 Apr.1975, p.2.
4. For an explanation of why this is described as 'inevitable' see P.R. Odell, *Oil and World Power*, 2nd ed., Penguin, London, 1972, esp. ch.2. See too C. Levinson, *Capital, Inflation, and the Multi-Nationals*, London, 1971, ch.9. Proposals for a 'European Company' being canvassed in the EEC might make a big difference in this field.
5. See *The Times*, 9 Jan.1975, p.19.
6. See *The Observer*, 19 May 1974, p.3.
7. This is referred to by Diane Morgan in 'Hard Labour', the *Guardian*, 20 Sept. 1973, p.13, and substantiated by my own conversations.
8. See *The Seaman* (the monthly journal of the National Union of Seamen), Jan. 1974, p.15.
9. *The Times*, 13 Apr.1974, p.1.
10. *The Times*, 27 Jan.1976, p.1.
11. A Report of the Norwegian Ministry of Social Affairs has apparently accused companies operating in the Norwegian sector of ignoring the law and has revealed that some drilling companies will not let crews join unions, that they ignore labour-permit regulations, and sometimes provided bad working conditions. See *The Oil and Gas Journal*, vol.73, no.29, Newsletter, 21 July 1975.
12. See *The Financial Times*, 21 May 1974.
13. See Keith Harper, 'A New Platform for the Union Men', the *Guardian*, 11 Feb.1975.
14. *The Times*, 5 June 1975, p.2.
15. A similar requirement is to be found in s.32 (1) of the Petroleum and

Submarine Pipe-lines Act 1975.
16. See the Mineral Workings (Offshore Installations) Act, s.6 and the Schedule, esp. paras.5, 6 and 8, and S.I. 1976 No.1019.
17. *H.L. Debates,* vol.315, col.1271, 2 Mar. 1971.
18. See the Employment Protection (Offshore Employment) Order 1976, S.I. 1976 No.766.
19. Ibid., ss.53 (1) (a) and 54 (1) and (3).
20. Trade Union and Labour Relations Act 1974, Schedule 1, paras. 4 (1) and 6 (4) (a). Note that as such a dismissal would be one for an 'inadmissible reason' by virtue of para.6 (6), the employee is not prevented from bringing a claim because of the short length of time for which he has been employed. See para.10 (a) and 11 (1).
21. See the Employment Protection Act 1975, s.126 (1), and the Trade Union and Labour Relations Act 1974, s.30 (1).
22. Employment Protection Act 1975, s.53 (1) (b).
23. Ibid., s.53 (4).
24. Defined in s.30 (1) of the Trade Union and Labour Relations Act 1974 (as amended by s.3 (3) of the 1976 Act). See the Employment Protection Act 1975, s.126 (1).
25. Employment Protection Act 1975, s.53 (5) (a).
26. Ibid., s.53 (5) (b).
27. Ibid., s.11.
28. Ibid., s.12 (1).
29. Ibid., s.12 (4), (5) and (6).
30. Ibid., s.16.
31. Ibid., s.126 (8).
32. Ibid., s.17.
33. See ibid., ss.19-21, esp. s.21 (3) (a).
34. See ibid., s.18 (1) (e).
35. Ibid., s.57.
36. Ibid., s.58.
37. See ibid., s.14.
38. Ibid., s.98.
39. Ibid., Schedule 1, paras.1, 2(a), 10 and 11. Note s.126 (8) of the Act which provides that: 'For the purposes of this Act it is immaterial whether the law (apart from this Act) which governs any person's contract of employment is the law of the United Kingdom, or any part of the United Kingdom, or not.'
40. It should be recalled that most of the companies operating offshore are members of either the North Sea chapter of the International Drilling Contractors' Association or the UK Offshore Operators' Association, though of course neither of these organisations carries on collective bargaining with the unions.
41. Employment Protection Act 1975, Schedule 1, para.4.
42. Ibid., Schedule 1, paras.1 and 2 (b).
43. Ibid., Schedule 1, para.5 (1).
44. Ibid., Schedule 1, para.5 (2).
45. See ibid., s.127 (1) (a) and the Wages Councils Act 1959, s.1 (2) (a).
46. Wages Councils Act 1959, ss.1 (1) (c) and 3, as amended by the Employment Protection Act 1975, s.89 (1).
47. Ibid., s.11, as amended by the Employment Protection Act 1975, s.89 (2) (a) and Schedule 7, Part 1.
48. I.e. before amendment by the Trade Union and Labour Relations (Amendment) Act 1976, s.1 (d).
49. See the Trade Union and Labour Relations (Amendment) Act 1976 s.3 (2).

50. See the Merchant Shipping Act 1970, ss.42 (2); 39 (1) and (3) (b); and 30 as amended by the Merchant Shipping Act 1974, s.19 (3). See further the Merchant Shipping (Disciplinary Offences) Regulations 1972, S.I. 1972 No.1294, and the Merchant Shipping (Seamen's Wages and Accounts) Regulations 1972, S.I. 1972 No.1700. Between them these provisions both give and circumscribe the seaman's right to strike. Note that breach of some of the above provisions might also lead to difficulties with respect to the word 'only' in s.13 (1) of the Trade Union and Labour Relations Act 1974, even as amended.
51. Problems relating to conflicts of laws are dealt with below at pp.120-5 (in relation to contract), and pp.154-6 (in relation to torts).
52. The special problems of loaned employees are dealt with in Chapter 6 at pp.152-4.
53. See the Contracts of Employment Act 1972, s.6.
54. There does not seem to be any reported decision on the interpretation of either 'wholly or mainly outside Great Britain' or 'ordinarily' under the Contracts of Employment Act. 'Wholly or mainly' fell to be interpreted in *Franklin* v. *The Gramophone Co. Ltd.* [1948] 1 K.B. 542, in which Somervell L.J. makes it clear (at pp.555 and 557) that the question is one of fact. In *Berthelemy* v. *Neale* [1952] 1 All E.R. 437 (C.A.) Sir Raymond Evershed M.R. said, in interpreting similar words in s.20 (1) of the Leasehold Property (Temporary Provisions) Act 1951, that the question involved 'common sense'. In *Re Hatschek's Patents, ex parte Zerenner* [1909] 2 Ch. 68, Parker J. had to interpret s.27 of the Patents and Designs Act 1907, which incorporated the phrase 'exclusively or mainly outside the United Kingdom'. He said (at p.83): 'There is no difficulty in the use of the word "exclusively", but the use of the word "mainly" gives rise to difficulty.' Adapting the test he then proposed, which related to manufactures, it would read:

    > 'I do not think that (employment) can be said to be mainly abroad merely because it is carried (out). . .abroad to a somewhat greater extent than in the United Kingdom. . .to come within the (definition) the disparity (between employment abroad and employment in this country) must, in my opinion, be greater than a small percentage and indeed. . .(must be). . . substantial.'

    On the other hand the Court of Appeal in *Miller* v. *Owners of the Ship Ottilie* [1944] K.B. 188 took the view, when interpreting another statute, that the word 'mainly' had the meaning given to it by the Oxford Dictionary: 'for the most part; chiefly; principally', and that this was factual. The decision is capable of interpretation in favour of a majority 'quantitative' approach.

    The word 'ordinarily' has fallen to be decided under s.1 (1) of the Maintenance Orders Act 1950. See, *inter alia; Macrae* v. *Macrae* [1949] 2 All E.R. 34 (per Somervell L.J. at p.36); and *Lewis* v. *Lewis* [1956] 1 All E.R. 375. In the later case Willmer J. said: 'I think myself it is a matter which has to be viewed in the context of all the other circumstances of the case.' (At p.378.) See too *R.* v. *Edgehill* [1963] 1 All E.R. 181.
55. In 1973, the time these contracts were being issued, the Contracts of Employment Act 1972, s.1, laid down the following periods of notice: 13 weeks to 2 years: 1 week; 2-5 years: 2 weeks; 5-10 years: 4 weeks; 10-15 years: 6 weeks; over 15 years: 8 weeks. Unders s.1 (5) and Schedule 1, para.1 (2) weeks in which an employee otherwise covered by the Act worked 'wholly or mainly outside Great Britain' were counted as full weeks of

employment.

The Employment Protection Act 1975, s.126 (5) and Schedule 16, Part 2, paras.1 and 2 have amended the provisions of the Contracts of Employment Act so that the minimum periods of notice are now: 4 weeks to 2 years: 1 week; from 2 years to 12 years: at least 1 week for every year of continuous employment; not less than twelve weeks thereafter. Unfortunately more recent contracts have not been available.

56. Now amended by the Employment Protection Act 1975, s.126 (5) and Schedule 16, Part 2, para.5.
57. *The Industrial Relations Code of Practice,* HMSO, London, 1972, was issued with the authority of Parliament, having been presented under s.3 (1) of the Industrial Relations Act 1971. It came into force on 28 Feb.1972. See the Industrial Relations (Code of Practice) Order 1972, S.I. 1972 No.179, and continues to be relevant in proceedings before industrial tribunals by virtue of the Trade Union and Labour Relations Act 1974, s.1 (2) (a) and Schedule 1, Part 1. See too the Employment Protection Act 1975, s.6.
58. BP installation managers were paid over £200 a week in 1973.
59. See the *Guardian,* 29 Aug.1973, p.7.
60. See *The Financial Times,* 7 Oct.1974.
61. *Note:* The figures quoted here are all 1973 scales. They will all have increased since then and in many cases the basis for calculation may have changed. It must also be added that obtaining reliable and accurate information relating to the wages of offshore workers is hard. Firstly there are variations. Secondly the companies will not reveal the figures, and finally it is difficult to find the workers themselves to check figures obtained. Those used in the text are those obtained by examination of the contracts in 1973. More recent figures are taken from various sources. Their accuracy is often in doubt. Overall it would seem that unskilled offshore workers are paid much less than many reports would have one believe.
62. These basic rates are supplemented by other payments considered below. Gross annual earnings may be between £2,500 and £3,000.
63. See the facts given in *Sayers* v. *International Drilling Co. N.V.,* [1971] 1 W.L.R. 1176. Sayers was to get £120 per month plus a monthly bonus of £35 and a platform allowance of £2 per day.
64. See William Allsopp, 'Unions Chase Tartan Coolies', *Sunday Times,* 30 June 1974, p.63; See too Keith Harper, 'A New Platform for the Union Men', the *Guardian,* 11 Feb.1975, p.13.
65. These are discussed separately in Chapter 7 at pp.180-4.
66. See Hepple & O'Higgins, ch.8, pp.67-74; Rideout, *Principles of Labour Law,* p.88 *et seq.;* A.E. Turner, *The Law of Trade Secrets,* Sweet & Maxwell, London, 1962; Gareth Jones, 'Restitution of Benefits Obtained in Breach of Another's Confidence', (1970) 86 L.Q.R. 463; and a series of articles by P.M. North: 'Disclosure of Confidential Information', [1965] J.B.L. 31; 'Further Disclosures of Confidential Information' [1968] J.B.L. 32; and 'Breach of Confidence: Is there a New Tort?' (1972) 12 J.S.P.T.L. 149.
67. As was the case in *Hivac Ltd.* v. *Park Royal Scientific Instruments* [1946] Ch. 169.
68. *Herbert Morris* v. *Saxelby* [1916] 1 A.C. 688.
69. Information sent from the drilling log is often 'scrambled' because of its value and the danger of industrial espionage. See *The Oil and Gas Journal,* vol.72, no.15, 18 Apr.1974, p.83.
70. *Merryweather* v. *Moore* [1892] 2 Ch. 518; *Bent's Brewery Co. Ltd.* v. *Hogan* [1945] 2 All E.R. 570. In *Hivac Ltd., etc.* (above) Lord Greene M.R. said at p.174: '. . .the obligation of fidelity, which is an implied term of the contract,

may extend very much further in the case of one class of employee than it does in others.'

71. *Printers & Finishers Ltd.* v. *Holloway* [1965] R.P.C. 239. But see *Allilueve* v. *Flegon, The Times,* 18 Aug.1967 per Geoffrey Lane J.
72. See *Fraser* v. *Evans* [1969] 1 Q.B. 349 per Lord Denning at p.361.
73. See *Stevenson, Jordan & Harrison Ltd.* v. *MacDonald & Evans* (1951) 68 R.P.C. 190, and (1952) 69 R.P.C. 10.
74. *Seager* v. *Copydex* [1967] 1 W.L.R. 923; *Saltman Engineering Co. Ltd.* v. *Campbell Engineering Co. Ltd.* (1948) 65 R.P.C. 203 per Lord Greene M.R. at p.213; *Terrapin Ltd.* v. *Builders Supply Co. (Hayes) Ltd.* [1960] R.P.C. 128 per Roxburgh J. at p.130; *Cranleigh Precision Engineering Ltd.* v. *Bryant* [1965] 1 W.L.R. 1293 per Roskill J. at pp.1317-19.
75. *Printers & Finishers* v. *Holloway* [1965] R.P.C. 239 per Cross J. at p.255.
76. See *Herbert Morris* v. *Saxelby* [1916] 1 A.C. 688.
77. See *Printers & Finishers* v. *Holloway* [1965] R.P.C. 239 per Cross J. at p.255: 'The mere fact that the information is not embodied in a document but is carried away by the employee in his head is not, of course, itself a reason against the granting of an injunction to prevent its use or disclosure by him.' This statement, and similar remarks in *Reid & Sigrist Ltd.* v. *Moss and Mechanism Ltd.* (1932) 49 R.P.C. 461 contradict the opinion of Kekewich J. in *Merryweather* v. *Moore* [1892] 2 Ch. 518. See too *Amber Size & Chemical Co. Ltd.* v. *Menzel* [1913] 2 Ch. 239.
78. See *Cranleigh Precision Engineering Ltd.* v. *Bryant* [1965] 1 W.L.R. 1293.
79. In journals such as *Petroleum Economist* and others.
80. This too is a subject best left to specialist works: see T.A. Blanco-White and R. Jacob, *Patents, Trade Marks, Copyright and Industrial Designs,* Sweet & Maxwell, London, 1970; T.A. Blanco-White, *Patents for Inventions,* 3rd ed., Stevens, London, 1962; D. Falconer, W. Aldous, and D. Young, *Terrell on the Law of Patents,* 12th ed., Sweet & Maxwell, London, 1971.
81. See *Selz Ltd.'s Application* (1953) 71 R.P.C. 158.
82. Under s.56 (2) of the Patents Act 1949, but see *Sterling Engineering Co. Ltd.* v. *Patchett* [1955] A.C. 534.
83. See A.D. Russell-Clarke *Copyright in Industrial Design,* 4th ed., Sweet & Maxwell, London, 1968; E.P. Skone James, *Copinger & Skone James on Copyright,* 11th ed., Sweet & Maxwell, London, 1971.
84. See *Sayers* v. *International Drilling Co. N.V.* [1971] 1 W.L.R. 1176 (C.A.). The relevant clause is reproduced below.
85. E.g. Hepple & O'Higgins, op.cit.
86. This would appear also to have been true of s.81 of the Employment Protection Act 1975, which gives the right to an itemised pay statement, even without extension of the Act, were payment made in Great Britain and it were not for s.119 (5). The section is not yet in force.
87. Counter-Inflation Act 1973, s.7 (3). See now the Prices Act 1974, and the Counter-Inflation (Abolition of Pay Board) Order 1974, S.I. 1974 No.1218.
88. Employment Protection Act 1975, ss.22-8, not yet in force.
89. Ibid., s.126 (8).
90. This section is concerned only with problems which are strictly a matter of contract. It is clear that many of the duties owed at common law, e.g. the duty ot take care, may also be seen as implied terms of the contract, and actionable either in tort or contract: *Matthews* v. *Kuwait Bechtel Corporation* [1959] 2 Q.B. 57 (C.A.); [1959] 2 All E.R. 345. See too *Davie* v. *New Merton Board Mills Ltd.* [1959] A.C. 604, at pp.619 and 642. Issues arising out of accidents and injuries (for example, those raised by *Sayers* v. *International Drilling Co. N.V.* [1971] 1 W.L.R. 1176; [1971] 3 All E.R.

163), are discussed in Chapter 6.

91. See, generally: G.C. Cheshire and P.M. North, *Cheshire's Private International Law,* 9th ed., Butterworths, London, 1974; J.H.C. Morris *Dicey & Morris: The Conflict of Laws,* 9th ed., Stevens, London, 1973; J.H.C. Morris, *The Conflict of Laws,* Stevens, London, 1971; and in particular: *Acts of the Second International Congress on Labour Law,* (Geneva, 12-14 Sept.1957), Geneva, 1961, pp.483-92 (O. Kahn-Freund); I. Szaszy, 'The Proper Law of Labour Contracts' (1968) 17 I.C.L.Q. 11; L.A. Collins, 'Interaction between Contract and Tort in the Conflict of Laws' (1967) 16 I.C.L.Q. 103. The subject has been covered by M.H. Van Hoogstraten in 'Conflict of Laws and Contracts of Employment', unpublished thesis submitted for the Diploma in Comparative Legal Studies, Cambridge, 1973.
92. *Not* 'which law governs the contract?', see Cheshire, op.cit., p.261. Cf. s.126 (8) of the Employment Protection Act 1975.
93. See *Re United Railways of the Havana and Regla Warehouses* [1960] Ch.52.
94. See *Kahler* v. *Midland Bank* [1950] A.C. 24, per Lord McDermott at p.42. The rule is known as 'decapage'. See Cheshire, op.cit., p.202.
95. This rule is of long standing: see *Gienar* v. *Meyer* (1796) 2 H. Bl. 603; 126 E.R. 728; *Vita Food Products Inc.* v. *Unus Shipping Co. Ltd.* [1939] A.C. 277 (P.C.); but see *Re Claim by Helbert Wagg & Co. Ltd.* [1956] Ch. 323 per Upjohn J. at p.34; and *Boissevain* v. *Weil* [1949] 1 K.B. 482 per Denning L.J. at p.491. See too *Companie D'Armement Maritime S.A.* v. *Companie Tunisienne de Navigation S.A.* [1971] A.C. 572.
96. See *Ocean Steamship Co.* v. *Queensland State Wheat Board* [1941] 1 K.B. 402.
97. See *The Assunzione* [1954] P. 150 per Singleton L.J. at pp.175, 179.
98. See *Tompkinson* v. *First Pennsylvania Banking and Trust Co.* [1961] A.C. 1007; *Philipson-Stow* v. *I.R.C.* [1961] A.C. 727 at p.760; *Companie D'Armement Maritime S.A.* v. *Companie Tunisienne de Navigation S.A., supra; James Miller & Partners Ltd.* v. *Whitworth Street Estates (Manchester) Ltd.* [1970] A.C. 583; *Coast Lines Ltd.* v. *Hudig and Veder Chartering N.V.* [1972] 2 Q.B. 34.
99. *Re Anglo-Austrian Bank* [1920] 1 Ch. 69. See too *Pick* v. *Manufacturers Life Insurance Co.* [1958] 2 Ll.R. 93; *Rossano* v. *Manufacturers Life Insurance Co.* [1963] 2 Q.B. 352.
100. See *Keiner* v. *Keiner* [1952] 1 All E.R. 643.
101. *South African Breweries* v. *King, infra; Re Hewitts Settlement, Hewitt* v. *Hewitt* [1915] 1 Ch. 228.
102. *Companie d'Armement Maritime S.A.* v. *Companie Tunisienne de Navigation S.A.* [1971] A.C. 572 at p.594; *Sayers* v. *International Drilling Co. N.C.* [1971] 1 W.L.R. 1176 at pp.1183-1184 and 1186; *Coast Lines Ltd.* v. *Hudig and Veder Chartering N.C.* [1972] 2 Q.B. 34 at pp.47 and 50.
103. [1899] 2 Ch. 173; affirmed [1900] 1 Ch. 273.
104. [1971] 1 W.L.R. 1176; [1971] 3 All E.R. 163.
105. [1971] 1 W.L.R. 1176 at p.1184 F.
106. His views on the 'law of the issue' are discussed in Chapter 6.
107. *Brodin* v. *A/R Seijan* 1973 S.L.T. 198.
108. *Acts of the Second International Congress on Labour Law,* p.486.
109. Ibid., p.485.

CHAPTER 5

1. John R. Huff (Manager, Engineering and Construction, Western Oceanic, Inc.),

'Rig Casualties – Statistics and Analysis', paper delivered at Offshore North Sea 1974 Conference, Stavenger, Norway, 3-6 Sept.1974, in the Safety and Environmental Protection Section.
2. E.g. the losses of the 'Transocean III', Jan.1975, and the 'Deep Sea Driller', Mar.1976.
3. See *Petroleum Times,* 11 Jan.1975, p.5; *H.C. Debates,* vol.867, cols.669-96, 16 Jan.1975. Huff, op.cit., estimates that up to April 1974 six losses in the North Sea cost $38.3 million.
4. *The Times,* 30 Nov.1972, p.1.
5. *The Times,* 21 Feb.1975, p.18.
6. See *The Times,* 8 Jan.1976, p.16.
7. *H.C. Debates,* vol.816, col.656 (Kevin McNamara), 28 Apr.1971.
8. *H.C. Debates,* vol.816, col.646 (Nicholas Ridley), 28 Apr.1971. See too *H.C. Debates,* vol.821, col.674, 14 July 1971.
9. *The Times,* 3 Nov.1975, p.1; *The Times,* 6 Nov.1975, p.3.
10. See *The Times,* 1 Sept.1975, p.3.
11. See *Development of the oil and gas resources of the U.K.,* 1975, p.7.
12. *The Times,* 16 Aug.1975, p.4.
13. *H.C. Debates,* vol.816, col.666 (John Prescott), 28 Apr.1971.
14. *H.L. Debates,* vol.315, col.742 (Earl Ferrers), 18 Feb.1971.
15. See the *'Sea Gem' Report,* para.1.10.
16. Colin Newsome, 'Weather poses a Tough Task for the Forecaster', *The Times,* 20 Oct.1972, p.15.
17. The Offshore Installations (Logbooks and Registration of Death) Regulations 1972, S.I. 1972 No.1542 (made under ss.6 and 7 of, and paras.11 and 12 of the Schedule to, the Act), Regulation 3 (2) (c).
18. Ibid., Regulation 6.
19. The Offshore Installations (Inspectors and Casualties) Regulations 1973, S.I. 1973 No.1842, Regulation 9.
20. Ibid., Regulation 12 (1).
21. See too *H.C. Debates,* Standing Committee G, *Official Report,* 1970/1 (Mineral Workings (Offshore Installations) Bill), cols.71-8, 22 June 1971; *H.C. Debates,* vol.818, written answers, cols.345-6, 9 June 1971.
22. These figures are all taken from *Development of the oil and gas resources of the U.K.,* 1975, pp.6 and 7 and Appendix 14.
23. See *The Times,* 11 Sept.1975, p.2. But see *The Times,* 20 Jan.1975.
24. See the *Report of the Working Party on the Medical Implications of Oil Related Industry,* BMA, Edinburgh, Sept.1975.
25. See *The Times,* 11 Sept.1975, p.2; *The Times,* 16 Sept.1975, p.3. See too *The Times,* 12 Sept. 1975, p.4.
26. CIRIA Underwater Engineering Group, *Principles of Safe Diving,* CIRIA, London, 1972.
27. For an analysis of the causes of some diving deaths see David Blundy, 'Pressures that kill the men who dive for Britain's oil', *The Sunday Times,* 23 June 1973, p.3.
28. *The Times,* 26 Aug.1975, p.1.
29. *The Times,* 12. Feb.1975, p.1.
30. *The Times,* 31 Oct.1975, p.14.
31. *The Times,* 25 Oct.1975, p.17.
32. Continental Shelf Act 1964, s.4 (1).
33. Coast Protection Act 1949, s.35 (1) (g). In relation to s.35 (1) (e) relating to the getting of minerals note the definition of 'surface' in s.49 (1).
34. See note 18 to Ch.2 above.
35. The Petroleum (Production) Regulations 1966, S.I. 1966 No.898, as amended. See now too the Offshore Installations (Operational Safety, Health and Welfare) Regulations 1976, S.I. 1976 No.1019, Regulation 7.

36. See p.48 above.
37. The words in brackets are added in the case of Exploration Licences.
38. *The Code of Safe Practice for Drilling, Production, and Pipeline Operations in Marine Areas, being Part VIII of the Institute of Petroleum Model Code of Safe Practice in the Petroleum Industry,* Institute of Petroleum, London, 1964. (Referred to hereafter as the *Code.)*
39. *Code,* Foreword.
40. The *'Sea Gem' Report,* para. 10.2 (1). See now the Mineral Workings (Offshore Installations) Act 1971, Schedule, para.15.
41. *Code,* para. 1.1.
42. *'Sea Gem' Report,* para.8.3.
43. According to the *'Sea Gem' Report,* para.8.3.
44. The Code also recommends that: 'Only persons who, in the opinion of the supervisor, are essential to the safe raising and lowering of a self-elevating type of mobile drilling platform, should be on the platform when this is being done.' *(Code,* para.1.2)

    The *'Sea Gem' Report* found that many shift workers, caterers and other workers ancillary to this operation were on board when the platform was being jacked down. See para.8.8 which describes the situation and criticises the above recommendation of the Code.
45. Though the Code recommends that regular escape drill exercises should be held (para.1.2), none had been held on the 'Sea Gem'. See the *'Sea Gem' Report,* para.8.9.
46. *'Sea Gem' Report,* para.2.1 (1).
47. See generally the Schedule to the Act. Para.15 of the Schedule reads: 'The Regulations may, in prescribing standards of safety, or imposing other requirements, refer to, and make obligations depend on, the provisions of any recognised industrial code of practice as for the time being in force.'
48. *H.C. Debates,* vol.816, col.647, 28 Apr.1971.
49. The Offshore Installations (Registration) Regulations 1972, S.I. 1972 No.702.
50. But see the Mineral Workings (Offshore Installations) Act 1971, s.12 (1) which defines 'owner' as: '. . .the person who has registered the installation. . . or, if there is no such person, the person for the time being having the management of the installation, or of its main structure' and note that although the obligation to nofify the Secretary of State of changes in the position and of the particulars of the installation is imposed upon the owner (S.I. 1972 No.702, Regulation 6), the details required for registration clearly contemplate that the person seeking to register the installation and the owner might be separate persons (see Regulation 5 (1) (a) and (b)). The Secretary of State is also given authority to refuse registration if it appears to him that the person applying for registration 'is not in a position to discharge the duties imposed by the owner by or under the Act' (Regulation 3 (2) (a)) which clearly indicates that a person other than the owner might have made the application.
51. S.I. 1972 No.702, Regulations 5 and 6.
52. Ibid., Regulation 10 (1).
53. Ibid., Regulation 9.
54. Ibid., Regulation 11.
55. See the *'Sea Gem' Report,* para.1.1.
56. Mineral Workings (Offshore Installations) Act 1971, s.3 (1) (a).
57. Ibid., s.3 (1) (b) and (c) and s.3 (2).
58. Ibid., s.3 (4).
59. The Offshore Installations (Construction and Survey) Regulations 1974, S.I. 1974 No.289

60. Department of Energy, *Guidance on the Design and Construction of Offshore Installations,* HMSO, London, 1974.
61. The Offshore Installations (Construction and Survey) Regulations 1974, S.I. 1974 No.289, Regulation 3 (1).
62. Ibid., Regulation 3 (2).
63. The Secretary of State has appointed Lloyd's Register of Shipping; American Bureau of Shipping; Bureau Veritas; Germanischer Lloyds; and Der Norske Veritas as competent surveying authorities. Reports that most of the work was going to the Norwegians *(The Times,* 29 May 1974) have been denied by Lloyd's (see *The Times,* 5 June 1974).
64. See the Offshore Installations (Construction and Survey) Regulations 1974, S.I. 1974 No.289, Schedule 2, Part 2.
65. Ibid., Regulation 5 (2).
66. Ibid., Regulation 6 (2).
67. Ibid., Regulation 5 (1) (d). See too S.I. 1976 No.1019, Regulations 2 and 4.
68. See ibid., Schedule 2, Parts 4 and 5.
69. See ibid., Schedule 2, Parts 6, 7 and 8.
70. Ibid., Regulation 8 (1).
71. Ibid., Regulation 7 (3) and (4).
72. Ibid., Regulation 11.
73. Ibid., Regulation 12.
74. Mineral Workings (Offshore Installations) Act 1971, s.6 (1).
75. Ibid., s.6 (2) (a).
76. Ibid., s.6 (2) (b).
77. Ibid., s.6 (2) (c).
78. Ibid., s.6 (2) (d).
79. See ibid., Schedule, para.3.
80. Ibid., Schedule, para.4 (1).
81. Some of the things mentioned in the *H.C. Debates* as being the subject of future regulations were foghorns, safety rockets, lights, divers, public address systems, radio contact, 'Brucker' capsules, and lifeboats.
82. In the Second Reading Committee Mr David Watkins (Consett) said;

> . . .I confirm what the Minister has said about work being in an advanced stage on the preparation of the regulations. In the course of a convention I had last week – I think this is almost classic and worth putting on record – I was told that the regulations would be ready 'in days rather than months'.

Mr. Ridley: 'We have heard that phrase before. We shall wait and see.' *(H.C. Debates,* vol.816, cols.674-5, 28 Apr.1971.)

83. The Offshore Installations (Inspectors and Casualties) Regulations 1973, S.I. 1973 No.1842, Part 1.

83a. By s.44 (5) (a) of the Petroleum and Submarine Pipe-lines Act 1975 the power to make regulations under s.6 of the 1971 Act is declared to include the power to make regulations providing for an inspector to enter upon any premises or vessel used or intended to be used in connection with an offshore installation, and to inspect them and anything on them which is used or intended to be used in such a manner. No amendment to the regulations has yet been made to give effect to this provision.

84. The Offshore Installations (Inspectors and Casualties) Regulations 1973, S.I. 1973 No.1842, Regulation 2 (1).
85. See ibid., Regulations 3 and 4.
86. See ibid., Regulations 7 and 8.

87. See ibid., Regulation 3 (1) (g).
88. Ibid., Regulation 3 (1) (e).
89. Ibid., Regulation 3 (1) (f).
90. The Offshore Installations (Logbooks and Registration of Death) Regulations 1972, S.I. 1972 No.1542 (made under the Mineral Workings (Offshore Installations) Act 1971, ss.6 and 7 and Schedule, paras.11 and 12), Regulation 2.
91. By ibid., Regulation 1 (3) they do not apply to installations registered as vessels which are dredging; installations in transit; or those unmanned. Regulation 2 allows the log book of a fixed installation under construction to be maintained on an attendant vessel.
92. See ibid., Regulation 7, and 3 (2) (a).
93. Ibid., Regulation 3 (3).
94. Ibid., Regulation 6.
95. Ibid., Regulation 3 (2).
96. On pain of a fine not exceeding £100 on summary conviction, see ibid., Regulation 12 (1).
97. Ibid., Regulation 3 (2) (b).
98. Ibid., Regulation 3 (2) (c).
99. Ibid., Regulation 3 (2) (d).
100. Ibid., Regulation 3 (2) (f).
102. The *'Sea Gem' Report,* para.10.2 (v).
103. The Offshore Installations (Inspectors and Casualties) Regulations 1973, S.I. 1973 No.1842, Part 2, and the Offshore Installations (Public Inquiries) Regulations 1974, S.I. 1974 No.338.
104. See pp.168-71.
105. See the Offshore Installations (Application of the Employers' Liability (Compulsory Insurance) Act 1969) Regulations 1975, S.I. 1975 No.1289, and the Employers' Liability (Compulsory Insurance) (Offshore Installations) Regulations 1975, S.I. 1975 No.1443. Discussed at pp.163-4 below.
106. The Offshore Installations (Diving Operations) Regulations 1974, S.I. 1974 No.1229. See Regulation 2 (1).
107. Ibid., Regulation 3 (5).
108. Ibid., Regulations 4 and 5.
109. Ibid., Regulation 6 (1) (a).
110. Ibid., Regulation 6 (1) (b).
111. Ibid., Regulation 6 (1) (c).
112. Ibid., Regulation 6 (1) (d).
113. Ibid., Regulation 6 (1) (e).
114. Ibid., Regulation 7 (1).
115. Ibid., Regulation 8 and Schedule 1.
116. Ibid., Regulation 10.
117. Ibid., Regulation 11.
118. Ibid., Regulation 9 (1).
119. Ibid., Regulation 9 (1) (e).
120. Ibid., Regulation 9 (1) (f).
121. Ibid., Regulation 9 (1) (g).
122. Ibid., Regulation 16.
123. Ibid., Regulation 17.
124. Ibid., Regulation 18.
125. Ibid., Regulation 15.
126. Ibid., Regulation 19.
127. Ibid., Regulation 20.
128. Ibid., Regulation 13.

129. Ibid., Regulation 12.
130. See the Merchant Shipping Act 1974, s.17 (3) and Schedule 5.
131. Ibid., s.16 (1). See too the Merchant Shipping (Registration of Submersible Craft) Regulations 1976, S.I. 1976 No.940.
132. The Merchant Shipping (Diving Operations) Regulations 1975, S.I. 1975 No.116, amended by the Merchant Shipping (Diving Operations) (Amendment) Regulations 1975, S.I. 1975 No.2062.
133. Ibid., Regulation 7 (c) to (e).
134. Ibid., Regulation 8.
135. Ibid., Regulations 4, 5 and 6.
136. Ibid., Regulations 21-4.
137. Petroleum and Submarine Pipe-lines Act 1975, s.26 (1). See pp.51-2 above.
138. See ibid., s.26 (2) reproduced at p.52 above.
139. Ibid., s.32 (1), and see the Submarine Pipe-lines (Diving Operations) Regulations 1976, S.I. 1976 No.923.
140. Ibid., s.32 (4).
141. Ibid., s.32 (5).
142. Ibid., s.27 (1).
143. Ibid., s.27 (2).
144. Ibid., s.27 (3).
145. Note that actions have already been brought for failure to observe the duties laid down by s.3 (1) of the Act in relation to accidents to divers working in Scapa Flow. See *The Times*, 10 Dec.1975, p.3. These occurred within territorial waters, however.
146. See the Health and Safety at Work etc. Act 1974, ss.1 (2), 15 (3), 53 and Schedule 1.
147. See ibid., s.84 (3) and s.53.
148. Ibid., s.84 (4) (b) and (c).
149. Ibid., s.80 (1) and (2) (a).
150. Ibid., s.15 (9).
151. *H.C. Debates,* vol.821, col.676 (John Prescott), 14 July 1971. See too *The Press and Journal,* Aberdeen, 28 Aug.1972, p.11. This renders the provisions of s.7 (2) (b) and (c) of the Sex Discrimination Act 1975 very relevant.
152. Shell, Conoco, Reading & Bates, and ODECO and many others, all ban alcohol on their rigs.

CHAPTER 6

1. Lord Wright in *Wilsons' & Clyde Coal Co, Ltd.* v. *English* [1938] A.C. 57 at p.78, quoting from Lord McLaren in *Bett* v. *Delmany* Coal Co. (1905) 7 F. (Ct. of Sess.) 787.
2. The whole subject is dealt with in detail in John Munkman, *Employers' Liability at Common Law,* 7th ed., Butterworths, London, 1971. For Scottish authorities see D.M. Walker, *The Law of Delicts in Scotland,* W. Green & Son, Edinburgh, 1966, pp.556-86.
3. See per Viscount Simonds in *Davie* v. *New Merton Board Mills Ltd.* [1959] 1 All E.R. 346 at p.350D.
4. [1959] 2 Q.B. 57 (C.A.).
5. *Paris* v. *Stepney Borough Council* [1951] A.C. 367 per Lord Morton. See too *Ellis* v. *Ocean S.S. Co. Ltd.* [1958] 2 Lloyd's Rep. 373 (C.A.).
6. *Saul* v. *St. Andrews Steam Fishing Co. Ltd.* [1965] 2 Lloyd's Rep. 107. For an example of a 'proper system of work' see *Speed* v. *Thomas Swift & Co. Ltd.* [1943] K.B. 557.
7. See *Street on Torts,* Butterworths, 5th ed., London, 1972, p.161.

8. [1944] K.B. 476.
9. Ibid., at pp.480-81.
10. [1965] A.C. 656.
11. *Wilsons' & etc., supra,* at pp.83-4 per Lord Wright.
12. Following the case of *Davie* v. *New Merton Board Mills Ltd.* [1959] A.C. 604, s.1 (1) of the Employers' Liability (Defective Equipment) Act 1969 stipulates that:

    Where after the commencement of this Act

    (a) an employee suffers personal injury in the course of his employment in consequence of a defect in equipment provided by his employer for the purpose of the employer's business; and
    (b) the defect is attributable wholly or partly to the fault of a third party (whether identified or not), the injury shall be deemed to be also attributable to negligence on the part of the employer (whether or not he is liable in respect of the injury apart from this subsection), but without prejudice to the law relating to contributory negligence. . .

    By s.1 (2) any agreement to exclude or limit the employer's liability under the Act is void.
13. There are innumerable cases, see Munkman, op.cit. Note too that safety consultants owe a duty of care to a contractor's employees: *Driver* v. *William Willett (Contractors) Ltd.* [1969] 1 All E.R. 665.
14. See Munkman, op.cit.; G. Weber (ed.), *Batt's Law of Master and Servant,* 5th ed.; Pitman's, London, 1967. For two cases concerning petroleum and naked lights see *Century Insurance Co.* v. *Northern Ireland Road Transport Board* [1942] A.C. 509; and *Dunk* v. *Hawker Saunders Ltd.* (1953), *The Times,* 27 Oct.1953, (C.A.),
15. [1945] 1 All E.R. 235.
16. Ibid., at p.237. See too *Lovell* v. *Blundells and T. & Albert Crompton & Co. Ltd.* [1944] K.B. 502 per Tucker J. at p.507.
17. Batt, op.cit., p.526.
18. Batt, op.cit., p.322 *et seq.*, and p.527.
19. [1924] A.C. 59.
20. Ibid., at p.71.
21. Cf. *R.* v. *National Insurance Commissioners, Ex parte East, The Times,* 1 Nov.1975, p.5.
22. *Rands* v. *McNeil* [1955] 1 Q.B. 235.
23. *Laszczyk* v. *The National Coal Board* [1954] 3 All E.R. 205.
24. *Smith* v. *Blandford Gee Cementation Co. Ltd.* [1970] 3 All E.R. 154, per Lord Parker C.J. at p.163g.
25. Including *Mersey Docks and Harbour Board* v. *Coggins & Griffiths (Liverpool) Ltd. and McFarlane* [1946] 2 All E.R. 345, [1947] A.C.1.
26. In *Denham* v. *Midland Employers' Mutual Assurance Ltd.* [1955] 2 All E.R. 561, at p.564; [1955] 2 Q.B. 437 at p.443.
27. See Munkman, op.cit., pp.83-5.
28. See Denning L.J. in *Denham* v. *Midland Employers' Mutual Assurance Ltd.* [1955] 2 All E.R. 561 at p.564I.
29. *Laugher* v. *Pointer* (1826) 5 B. & C. 547 per Littledale J. at p.558; 108 E.R. 204 at p.208. See also *O'Reilly* v. *Imperial Chemical Industries Ltd.* [1955] 1 W.L.R. 1155 per Parker C.J. at p.1162. In *Savory* v. *Holland, Hannen and Cubitts (Southern) Ltd.* [1964] 3 All E.R. 18, Lord Denning M.R. said that the latter case was no longer an authority (at p.21).

30. [1974] A.C.1. See per Lord Porter at p.17.
31. See *Mersey Docks etc.*, per Viscount Simon at p.10.
32. *Savory* v. *Holland, Hannen and Cubitts (Southern) Ltd.* [1964] 3 All E.R. 18. *O'Reilly* v. *I.C.I. Ltd.* [1953] 3 All E.R. 382.
33. *Garrard* v. *A.E. Southey & Co. and Standard Telephones and Cables Ltd.* [1952] 2 Q.B. 174.
34. *Denham* v. *Midland Employers' Mutual Assurance Ltd.*, op.cit.
35. [1959] 2 Q.B. 57 (C.A.). See too *Davie* v. *New Merton Board Mills Ltd.* [1966] 2 Q.B. 370, and *Sole* v. *W.J. Hallt Ltd.*, [1973] 1 Q.B. 574.
36. See generally Cheshire, op.cit., ch.10; *Dicey & Morris,* op.cit., ch.30 in Part 5; Morris, op.cit., ch.13.
37. As the situation was before the decision in *Chaplin* v. *Boys* [1971] A.C. 356, it was apparently adopted only by China and Japan outside the Commonwealth, see Lorenzen, *Selected Articles on the Conflict of Laws,* Yale U.P., New Haven, 1947, p.376.
38. (1870) L.R.6 Q.B.1. The rule is applied in Scotland: *M'Elroy* v. *M'Allister* (1949) S.C.110.
39. [1971] A.C.356.
40. [1971] 1 W.L.R. 1176 (C.A.).
41. See in particular: R. Smith, 'International Employment Contracts – Contracting Out' [1972] 21 I.C.L.Q. 164; L.A. Collins, 'Exemption Clauses, Employment Contracts and the Conflict of Laws' [1971] 20 I.C.L.Q. 320. See too L.J. Kovats, 'International Contracts of Employment' (1971) 121 N.L.J. 734.
42. See *Brodin* v. *A/R Seijan* 1973 S.L.T. 198 esp. at p.201.
43. *Brodin* v. *A/R Seijan* 1973 S.L.T. 198 in which *Sayer's* case was criticised and distinguished.
44. See Lord Wright in *London Passenger Transport Board* v. *Upson* [1949] A.C. 155 at p.168.
45. Not on fellow workers. See the distinction drawn in *Harrison* v. *National Coal Board* [1951] A.C.639.
46. See *Pasmore* v. *Oswaldtwistle Urban District Council* [1898] A.C. 387 per Lord Halsbury L.C., at p.397.
47. See the Mineral Workings (Offshore Installations) Act 1971, ss.3 (4), 4 (5) and 5 (4).
48. Ibid., s.5 (4).
49. See the Offshore Installations (Diving Operations) Regulations 1974, S.I. 1974 No.1229, Regulation 22.
50. See *Lochgelly Iron and Coal Co. Ltd.* v. *M'Mullan* [1934] A.C.1 per Lord Warrington at p.13. Section 11 (5) of the Mineral Workings (Offshore Installations) Act provides that the Act shall also bind the Crown wherever any of the provisions relate to a concession owner, in which case, for the purpose of the Act and the regulations, 'persons in the service of the Crown shall be taken to be employed whether or not they would be so taken apart from this subsection'. This latter statement reflects the uncertainty surrounding the legal position of industrial civil servants, particularly in relation to whether or not they have a contract of employment. The Act does not give rights of action to members of the armed services, and it does not bind the Sovereign in her personal capacity (s.11 (6) CI, s.30 (4) of the Petroleum and Submarine Pipe-Lines Act.
51. See *Hosking* v. *De Havilland Aircraft Co. Ltd.* [1949] 1 All E.R. 540; *Braham* v. *J. Lyons & Co. Ltd.* [1962] 3 All E.R. 281 (C.A.).
52. See *Grant* v. *National Coal Board* [1956] A.C. 649; *Donaghey* v. *Boulton &*

Paul Ltd. [1968] A.C.1; *Millard* v. *Serck Tubes Ltd.* [1969] 1 All E.R. 598 (C.A.).

53. E.g. *Cummings (or McWilliams)* v. *Sir William Arrol & Co. Ltd.* [1962] 1 All E.R. 623 (II.L)
54. Such as the preamble to the Limitation Act 1963 and the definition of personal injury given s.3 of the Law Reform (Personal Injuries) Act 1948. Cf. the Petroleum and Submarine Pipe-lines Act, s.30 (5).
55. Regulations could be made on these subjects under s.6 of, and para.5 of the Schedule to, the Mineral Workings (Offshore Installations) Act 1971.
56. Discussed below at p.178.
57. The words of s.9 (3) (b).
58. Occupiers' Liability Act 1957, s.1. The Act does not apply to Scotland: see s.8 (2), but see the Occupiers' Liability (Scotland) Act 1960. A discussion of the position in Scotland which highlights some of the differences, is found in D.M. Walker, op.cit., pp.586-602.
59. There is of course a distinction between the two. See generally P.M. North, *Occupiers' Liability*, Butterworths, London, 1971, ch.4.
60. *Wheat* v. *E. Lacon & Co. Ltd.* [1966] A.C. 552; *Fisher* v. *C.H.T. Ltd.* [1966] 2 Q.B. 475; for the position of sub-contractors see *Kearney* v. *Eric Waller, Ltd.* [1967] 1 Q.B. 29; *Bunker* v. *Charles Brand & Son Ltd.* [1969] 2 Q.B. 480.
61. [1943] 2 All E.R. 621 at p.627.
62. See *Cavalier* v. *Pope* [1906] A.C. 428.
63. See *Duncan* v. *Cammell Laird & Co. Ltd.* [1943] 2 All E.R. 621; *Creed* v. *John McGeoch & Sons Ltd.* [1955] 1 W.L.R. 1005; *Savory* v. *Holland, Hannen and Cubitts (Southern) Ltd.* [1964] 3 All E.R. 18; *Hartwell* v. *Grayson, Rollo and Clover Docks Ltd.* [1947] K.B. 901.
64. *Cavalier* v. *Pope* [1906] 4 C. 428; *Duncan* v. *Cammell Laird & Co. Ltd.* [1943] 2 All E.R. 621; *Hartwell* v. *Grayson, Rollo and Clover Docks Ltd.* [1947] K.B. 901 at p.914 . *Excelsior Wire Rope Co.* v. *Callan* [1930] A.C. 404; a number of Scottish cases have approved the same principle, see: *McIlwaine* v. *Stewarts Trustees,* 1914, S.C. 934; *Murdock* v. *A. and R. Scott,* 1956, S.C. 309.
65. Workmen employed by contractors or sub-contractors for the ultimate benefit of the occupier are his 'invitees', and thus his 'visitors' for the purposes of the Act. See *Wilkinson* v. *Rea Ltd.* [1941] 1 K.B. 688; *London Graving Dock* v. *Horton* [1951] A.C. 737.
66. Occupiers' Liability Act 1957, s.2 (6).
67. Ibid., s.2 (2).
68. For two analogous cases see: *Lee* v. *Blue Star Line* [1962] 1 Lloyd's Rep. 440; and *Gerrard* v. *Blue Star Line* [1962] 2 Lloyd's Rep. 294. Cf. *Bird* v. *King Line Ltd.* [1970] 1 Lloyd's Rep. 349.
69. For a discussion of the phrase 'without more' see F.H. Newark, 'The Occupiers' Liability Act (Northern Ireland) 1957', (1958) 12 N.I.L.Q. 203 at pp.211-13.
70. E.g. Conoco.
71. Occupiers' Liability Act 1957, s.2 (5).
72. *Third Report of the Law Reform Committee (Occupiers' Liability)*, Nov, 1954, Cmnd. 9305, para. 78 (vii) (x).
73. *London Graving Dock* v. *Horton* [1951] A.C. 737. See too *Bunker* v. *Charles Brand & Son Ltd.* [1969] 2 Q.B. 480.
74. See the Occupiers' Liability Act 1957, s.2 (1). The Occupiers' Liability (Scotland) Act 1960, s.2 (1) permits exclusion of liability by agreement only.
75. Cf. *Ashdown* v. *Samuel Williams & Sons Ltd.* [1957] 1 Q.B. 409 (C.A.). See too *White* v. *Blackmore* [1972] 3 W.L.R. 296 (C.A.), and *Bishop* v.

J.S. Starnes & Sons Ltd. [1971] 1 Lloyd's Rep. 162.

76. See the Occupiers Liability Act 1957, s.3.
77. See ibid., s.5.
78. There is no equivalent in the Scottish Act.
79. *Haseldine* v. *C.A. Daw & Son Ltd.* [1941] 2 K.B. 343 (C.A.); *Woodward* v. *The Mayor of Hastings* [1945] K.B. 174; but see Lord Reid in *Davie* v. *New Merton Board Mills Ltd.* [1959] A.C. 604 (H.L.) at p.644. The delegation in cases of the xpertise would appear to remain reasonable, see *Riden* v. *A.C. Billings & Sons Ltd.* [1957] 1 Q.B. 46 and *Green* v. *Fibreglass Ltd.* [1958] 2 Q.B. 245.
80. See the Fatal Accidents Act 1976, s.1. See too the Limitation Act 1975.
81. I.e. the section applies s.11 (1) to the Fatal Accidents Act.
82. Law Reform (Miscellaneous Provisions) Act 1934, s.1.
83. Schedule 1 of the Act sets out the Convention. Reference is to Article 17.
84. The carrier is not liable if he proves that he and his servants or agents have taken all necessary measures to avoid the damage or that it was impossible for him or them to take such measures, Article 20. Professor Street is quite plainly wrong in his book on torts about the application of the Act and its relation with the Fatal Accidents legislation, see *Street on Torts,* 5th ed., 1972, p.413.
85. The limitation period mentioned under Article 29 is two years. In relation to the limitation period generally see now the Limitation Act 1975.
86. About £6,000.
87. By Article 22. In fact the carrier cannot limit his liability if it is proved that the damage resulted from an act or omission of the carrier his servants or agents, done with intent to cause damage or recklessly and with knowledge that damage would probably result: Article 25.
88. See the Law Reform (Contributory Negligence) Act 1945, s.1.
89. Civil Aviation Act 1949, ss.40 (2) and 63 (3). This only applies to British Aircraft, see s.59.
90. Employers' Liability (Compulsory Insurance) Act 1969, s.1 (1).
91. The Offshore Installations (Application of the Employers' Liability (Compulsory Insurance) Act 1969) Regulation 1975, S.I. 1975 No.1289, Regulation 2. Made under the Mineral Workings (Offshore Installations) Act 1971, ss.6 and 7 and the Schedule, para.4 (2) (b). Note that Regulation 4 renders s.2 (2) (b) of the 1969 Act nugatory.
92. Ibid., Regulation 1 (2).
93. Ibid., Regulation 3 (a).
94. Ibid., Regulation 1 (2).
95. Ibid., Regulation 6.
96. The Employers' Liability (Compulsory Insurance) (Offshore Installations) Regulations 1975, S.I. 1975 No.1443, esp. Regulation 2 (c) which applies the Employers' Liability (Compulsory Insurance) General Regulations 1971, S.I. 1971 No.1117, as amended by S.I. 1974 No.208, with modifications. Regulation 3 applies the Employers' Liability (Compulsory Insurance) Exemption Regulations 1971, S.I. 1971 No.1933, as amended by S.I. 1974 No.208.
97. See the Continental Shelf Act 1964, s.10 (1), repealed by the Statute Law Revision (Consequential Repeals) Act 1965 as a result of s.76 of the National Insurance (Industrial Injuries) Act 1965.
98. The Social Security (Employed Earners' Employments for Industrial Injuries Purposes) Regulations 1975, S.I. 1975 No.467, Regulation 2 and Schedule 1, Part 1, para.7.
99. The Social Security Benefit (Persons Abroad) Regulations 1975, S.A. 1975

No.563, Regulation 11 (3). This would not apply to visitors, of course.
100. Social Security Act 1975, s.52.
101. See ibid., s.53.
102. See ibid., s.53 (1) (a).
103. Ibid., s.54.
104. The Social Security Benefit (Persons Abroad) Regulations 1975, Regulation 11 (4).
105. Social Security Act 1975, s.142. See too the Social Security (Reciprocal Agreements) Order 1975, S.I. 1975 No.813, Article 2 (1) and the Schedule, which lists Orders relating to all the other countries bordering on the North Sea.
106. The Social Security (Contributions) Regulations 1975, S.I. 1975 No.492, Regulations 76 (1).
107. Ibid., Regulation 76 (2).
108. Ibid., Regulation 76 (3).
109. The Social Security Benefits (Persons Abroad) Regulations 1975, S.I. 1975 No. 563, Regulations 11 (3).
110. Ibid., Regulation 11 (4).
111. See p.128 above.
112. The Offshore Installations (Inspectors and Casualties) Regulations 1973, S.I. 1973 No.1842, Regulation 9 (a).
113. Ibid., Regulation 9 (c) and the Schedule.
114. Ibid., Regulation 10.
115. See the Offshore Installations (Logbooks and Registration of Death) Regulations 1972, S.I. 1972 No.1542, Regulations 8 and 9. A return of death is not invalid on the grounds only that it is made outside the period laid down in the regulations.
116. 'Presumably' because the regulations do not say so specifically. It would seem implicit.
117. The Offshore Installations (Logbooks and Registration of Death) Regulations 1972, S.I. 1972 No.1542, Regulation 10.
118. The Offshore Installations (Inspectors and Casualties) Regulations 1973, S.I. 1973 No. 1842, Regulation 12 (1).
119. Ibid., Regulation 12 (2).
120. Ibid., Regulation 11.
121. See the *'Sea Gem' Report*, p.1.
122. *H.C. Debates*, vol.816; col.663 (David Watkins, MP for Consett), 28 Apr. 1971.
123. See the *'Sea Gem' Report*, p.1.
124. Kevin McNamara, MP for Hull North, *H.C. Debates*, vol.816, col.658, 28 Apr. 1971.
125. See the Mineral Workings (Offshore Installations) Act, 1971, Schedule, paras.10 (1) (b), (2) and (3).
126. The Offshore Installations (Public Inquiries) Regulations 1974, S.I. 1974 No.338, Regulations 2 and 3 (1).
127. Ibid., Regulation 4.
128. Ibid., Regulation 4 proviso.
129. Ibid., Regulation 5.
130. Ibid., Regulation 6 (a) to (c).
131. Ibid., Regulation 6 (d).
132. Ibid., Regulation 7.
133. Ibid., Regulation 10.
134. Ibid., Regulation 9 (1).
135. Ibid., Regulation 9 (2).

136. Ibid., Regulation 9 (3).
137. See the Coroners Act 1887, s.3 as modified by the Coroners (Amendment) Act 1926, s.21. See generally the Coroners Acts 1887-1954 as amended in particular by the Courts Act 1971 and the Local Government Act 1972.
138. See the Continental Shelf (Jurisdiction) (Amendment) Order 1975, S.I. 1975 No.1708.
139. Fatal Accidents and Sudden Deaths Inquiry (Scotland) Act 1976 s.1 (1) (a) (i).
140. Ibid., s.8.
141. Ibid., ss.2, 3 (2), 4 and 5 (1).
142. Ibid., s.6 (1) and (2).
143. Ibid., s.6 (3).
144. Mineral Workings (Offshore Installations) Act 1971, s.6 (5), added by ibid., s.8 (1) and Schedule 1, para.3.
145. Petroleum and Submarine Pipe-lines Act 1975, s.27 (3).
146. Ibid., s.27 (3) (a) and (b).
147. Ibid., s.27 (4) (b).
148. Ibid., s.27 (5) as amended by the Fatal Accidents and Sudden Deaths Inquiry (Scotland) Act 1976, s.8 (1) and Schedule 1, para.5.

CHAPTER 7

1. *'Sea Gem' Report*, para.10.2 (ii).
2. Ibid., para.10.2 (iii). See too para.8.4.
3. Ibid., para.10.1.
4. See *H.C. Debates*, vol.821, col.669 (Prescott and McNamara), 14 July 1971.
5. Some indication of the actions which will fall under the heading of safety can be gleaned from the subject matter of the regulations made under s.6 and the Schedule to the Act. Clearly the word is capable of a very wide definition, and the words 'health and welfare', particularly 'welfare' could be subject to the same treatment. Hours, rest periods, noise, control over leisure activities, and even clothing, could all be brought under these headings.
6. See *H.C. Debates*, vol.816, col.673 (Nicholas Ridley), 28 Apr. 1971.
7. Merchant Shipping Act 1970, s.79.
8. Mineral Workings (Offshore Installations) Act 1971, s.5 (8).
9. Ibid., s.5 (9).
10. Proceedings instituted in England alternatively need the consent of the DPP unless they already require the consent of the Attorney-General. In Northern Ireland they need the consent of the Attorney-General for Northern Ireland. S.10 (6) obviates the need for consent under s.3 of the Territorial Waters Jurisdiction Act 1878. See too the Health and Safety at Work etc. Act 1974, s.84 (4) (d), s.127 (3) (e) and (f) of the Employment Protection Act 1975, and s.29 (6) of the Petroleum and Submarine Pipe-lines Act 1975.
11. See *H.L. Debates*, vol.315, col.750 (Earl Ferrers, the Government spokesman), 18 Feb. 1971.
12. Mineral Workings (Offshore Installations) Act 1971, s.10 (5).
13. Note s.44 (1) of the Petroleum and Submarine Pipe-lines Act 1975 in relation to s.3 (4) of the Mineral Workings Act (Offshore Installations) Act 1971.
14. *H.C. Debates*, vol.821, col.673, 14 July 1971. See too S.I. 1976 No.1019, Reg.34.
15. Similar power in relation to regulations made under s.26 of the Petroleum and Submarine Pipe-lines Act 1975 is found in s.32 (3) of that Act.
16. Mineral Workings (Offshore Installations) Act 1971, s.7 (4), (5), (6) and (7). Cf. s.32 (4) of the Petroleum and Submarine Pipe-lines Act 1975.
17. Ibid., s.7 (8). Cf. s.46 of the Petroleum and Submarine Pipe-lines Act 1975.
18. Laid down in ibid., s.7 (1). It would seem to belie the demand in the *'Sea Gem' Report* for an 'accepted' discipline.

19. *H.C. Debates,* vol.821, 672 (moved by Eric Ogden), 14 July 1971.
20. See the reply given to the amendment by Nicholas Ridley at ibid., col.673-4, 14 July 1971.
21. The same wording is used in s.29 (4) of the Petroleum and Submarine Pipe-lines Act 1975 in relation to offences under that Act and any regulations made under it.
22. A repetition of similar wording in s.8 (3). See Chapter 2.
23. John Prescott MP., *H.C. Debates,* vol.816, co.672, 28 Apr. 1971. The refusal to 'fly-off' a man who demands to go ashore would in certain cases be justified, so a claim for wrongful imprisonment would fail. See *Herd* v. *Weardale Steel, Coke and Coal Co.* [1915] A.C. 67 (H.L.) for a good analogy. Cf. *Robinson* v. *Balmain New Ferry Co. Ltd.* [1910] A.C. 295 (P.C.).
24. *The Industrial Relations Code of Practice,* HMSO, London, 1972, paras.130-3 advises management to make known to employees the disciplinary rules and procedures. The Code is issued with the authority of Parliament, having been presented under s.3 of the Industrial Relations Act 1971, and remains relevant in proceedings before industrial tribunals by virtue of the Trade Union and Labour Relations Act 1974, s.1 (2) and Schedule 1, Part 1.
25. Ibid., para.133 (i).
26. Ibid., para 133 (ii).
27. See *H.C. Debates,* col.671, 14 July 1971.
28. *Court of Inquiry (Industrial Courts Act 1919) into certain matters concerning the shipping industry. Final Report,* (Chairman: Lord Pearson) Cmnd. 3211, HMSO, London 1967, para.287.
29. Ibid., para.289.
30. See E. Goffman, *Asylums – Essays on the Social Situation of Mental Patients and Other Inmates,* Doubleday Anchor, New York, 1961, Introduction.
31. See for example Vilhelm Aubert, *Elements of Sociology,* Heinemann, London, 1968, pp.170-82; Bryan Nolan and W.H. Hopwood, in P.H. Fricke (ed.), *Seafarer and Community,* Croom Helm, London, 1973, pp.85 and 97-116. For a good critical analysis see N. Perry and R. Wilkie, 'Social Theory and Shipboard Structure', Part 1: 'Some Reservations on an Emerging Orthodoxy', 1 *Maritime Studies and Management* 31, 1973; Part 2: 'Models, Metaphors and Ships', 2 *Maritime Studies and Management* 136, 1974.
32. Goffman, *supra,* p.9.
33. See *'Sea Gem' Report.*
34. See Pearson, para.327.
35. Trade Union and Labour Relations Act 1974, Schedule 1, para.9 (2). See p.63 ante.
36. See *Wales* v. *International Drilling Co. Ltd.* [1975] I.R.L.R. 341; *K. Melgaard* v. *Gurtler, Herbert & Co. Inc.* (unreported decision of Bury St. Edmunds tribunal, 6 June 1975, Case No.1031/75).
37. Trade Union and Labour Relations Act 1975, Schedule 1, para.9 (3).
38. Note however that the computation of the period of employment for the purposes of s.8 (1) of the Act as affected in either case by the provisions of subsections (3) to (6) of s.17 as amended by s.125 (1) of the Employment Protection Act 1975 and Schedule 16, Part 1, para.8.
39. Redundancy Payments Act 1965, s.1 (1) as amended by s.125 (1) of the Employment Protection Act 1975 and Schedule 16, Part 1, para.5 (1).
40. Employment Protection Act 1975, ss.99-107.
41. Ibid., s.101 (7).
42. Ibid., s.70.

INDEX TO STATUTES

A. STATUTES OF THE UNITED KINGDOM PARLIAMENT

(References are to the page on which the Act or section is discussed, and, following the oblique, to the note to that page where it is cited.)

B. FOREIGN LEGISLATION

Belgium

Denmark

United States

TABLE OF STATUTORY INSTRUMENTS

1964	No. 697	Continental Shelf (Designation of Areas) Order	46/150
1965	No.1531	Continental Shelf (Designation of Additional Areas) Order	46/150
1966	No. 898	Petroleum (Production) Regulations	47/158; 47/162; 132/35
1967	No. 648	Industrial Training (Petroleum Board) Order	77/19
	No. 655	Continental Shelf (Protection of Installations) Order (No. 1)	34/18
1968	No. 323	Continental Shelf (Protection of Installations) Order (No. 2)	34/18
	No. 891	Continental Shelf (Designation of Additional Areas) Order	46/150
	No. 892	Continental Shelf (Jurisdiction) Order	56/212; 57/214; 57/215-6
1969	No. 195	Continental Shelf (Protection of Installations) Order (No. 3)	34/18
	No. 1322	Continental Shelf (Protection of Installations) Order (No. 4)	34/18
1970	No. 205	Industrial Training (Petroleum Board) (Amendment) Order	77/19
	No. 1537	Secretary of State for Trade and Industry Order	46/155
1971	No. 594	Continental Shelf (Designation of Additional Areas) Order	46/150
	No. 721	Continental Shelf (Jurisdiction) (Amendment) Order	57/214
	No. 814	Petroleum (Production) (Amendment) Regulations	47/159; 47/161
	No. 1117	Employers' Liability (Compulsory Insurance) General Regulations	164/96
	No. 1933	Employers' Liability (Compulsory Insurance) Exemption Regulations	164/96
1972	No. 179	Industrial Relations (Code of Practice) Order	111/57

	No. 138	Ships and Mobile Offshore Installations Construction Credits (Increase of Limit) Order	25/84
	No. 467	Social Security (Employed Earners' Employment for Industrial Injuries Purposes) Regulations	165/98
	No. 492	Social Security (Contributions) Regulations	166/106-8
	No. 563	Social Security Benefit (Persons Abroad) Regulations	165/99; 166/104; 167/109-10
	No. 815	Social Security (Reciprocal Agreements) Order	166/105
	No. 1289	Offshore Installations (Application of the Employers' Liability (Compulsory Insurance) Act 1969) Regulations	142/105; 163-4/91-5
	No. 1443	Employers' Liability (Compulsory Insurance) (Offshore Installations) Regulations	142/105; 164/96
	No. 1708	Continental Shelf (Jurisdiction) (Amendment) Order	57/214; 170/138
	No. 1845	Sex Discrimination Act 1975 (Commencement) Order	84/40
	No. 2062	Merchant Shipping (Diving Operations) (Amendment) Regulations	144/132-6
	No. 2079	Petroleum and Submarine Pipe-lines Act 1975 (Commencement) Order	29/116
1976	No. 276	Petroleum (Production) Regulations	47/159
	No. 332	Continental Shelf (Protection of Installations) Order	34/18
	No. 766	Employment Protection (Offshore Employment) Order	66/245; 77/19; 103/18; 185-6
	No. 923	Submarine Pipe-lines (Diving Operations) Regulations	145/139
	No. 940	Merchant Shipping (Registration of Submersible Craft) Regulations	144/131
1976	No. 954	Continental Shelf (Protection of Installations) Order (no. 2)	34 & fn. 18

TABLE OF CASES

(References are to the page on which the case is discussed and, following the oblique, to the footnote reference on that page).

INDEX